A HISTORY OF ENGINEERING IN CLASSICAL AND MEDIEVAL TIMES

A HISTORY OF ENGINEERING IN CLASSICAL AND MEDIEVAL TIMES

Donald Hill

OPEN COURT PUBLISHING COMPANY

La Salle, Illinois

OPEN COURT and the above logo are registered in the US Patent and
Trademark Offices
OC845   10   9   8   7   6   5   4   3   2
Published by arrangement with Croom Helm Ltd, Beckenham, Kent

Library of Congress Cataloging in Publication Data

Hill, Donald Routledge
A history of engineering in classical and medieval times

   Bibliography : p. 248.
   Includes index.
   1. Enginering—Europe—History.
   2. Enginering—Asia—History.
   I. Title.
TA16.   H55   1984   620'.009'01   84-7339
ISBN 0-87548-422-0

Printed and bound in Great Britain

Contents

List of Plates   vi
List of Figures   vii
Abbreviations   xi
Acknowledgements   xii
Preface   xiii
 1. Introduction   1

Part One: Civil Engineering
 2. Irrigation and Water Supply   17
 3. Dams   47
 4. Bridges   61
 5. Roads   76
 6. Building Construction   98
 7. Surveying   116

Part Two: Mechanical Engineering
 8. Water-raising Machines   127
 9. Power from Water and Wind   155

Part Three: Fine Technology
10. Instruments   183
11. Automata   199
12. Clocks   223
Bibliography   248
Index   254

List of Plates

1. Cornalvo Dam   90
2. Muslim Weir at Valencia   91
3. Murcia Dam   92
4. Almonacid de Cuba Dam   93
5. Almansa Dam   94
6. Aqueduct Bridge — Aqua Claudia   95
7. Masonry of Aqua Claudia Bridge   96
8. Aqueduct Bridge — Mérida   97

List of Figures

1.1	Cultural Divisions	2
2.1	Hydraulic Map of Medieval Iraq	23
2.2	Qanat	35
2.3	Inverted Siphon	38
2.4	Roman Water-supply System	39
3.1	Hydraulic Works at Marib	51
4.1	Bridge Types	62
4.2	Types of Arch	67
5.1	Persian Road System	78
5.2	Cross-sections of Roman Roads	83
6.1	Gothic Building	100
6.2	Lifting Devices	110
6.3	Sheerlegs	111
7.1	Surveying Instruments	118
7.2	Groma	119
7.3	Back of Astrolabe	121
8.1	Shaduf	130
8.2	Screw	132
8.3	Tympanum	134
8.4	Saqiya	137
8.5	Spiral Scoop-wheel	138
8.6	Noria	140
8.7	Detail of Noria	141
8.8	Force-pump	143
8.9	Al-Jazarī's First Water-raising Machine	147
8.10	Al-Jazarī's Fourth Water-raising Machine	149
8.11	Detail of Figure 8.10	150
8.12	Al-Jazarī's Double-acting Pump	151
8.13	Detail of Figure 8.12	152
9.1	Undershot Water-wheel	156
9.2	Overshot Water-wheel	156
9.3	Gears for Mills	157

9.4	Horizontal Water-wheel	158
9.5	Windmill	174
10.1	Ptolemaic System	185
10.2	Antikythera Mechanism	187
10.3	Al-Bīrūnī's Calendar	189
10.4	Astrolabe	191
11.1a	Concentric Siphon	209
11.1b	Double Concentric Siphon	210
11.2	Automaton from Philo	214
11.3	Hero's Aeolipile	215
11.4	Lamp from Bānū Mūsà	216
11.5	Trick Vessel from Bānū Mūsà	217
11.6	Phlebotomy Measuring Device from al-Jazarī	218
11.7	Fountain from al-Jazarī	221
12.1	Ctesibius' Water-clock	228
12.2	Water-machinery from 'Archimedes' Clock	229
12.3	Face of Ridwān's Clock	234
12.4	Water-machinery from al-Jazarī's Clocks	237
12.5	Candle Clock from al-Jazarī	239
12.6	Verge Escapement	243

To the memory of
Rose and Henry Hill

Abbreviations

Only two abbreviations are used:

1. BGA means *Biblioteca geographorum arabicorum*. This is the title of a series of volumes of the works of Arab geographers published by Brill of Leiden. In the Notes and Bibliography, the details of individual volumes are given.

2. EI is used for *The Encyclopaedia of Islam*, 2nd edition, published jointly by E.J. Brill of Leiden and Luzac & Co., of London. The author, volume number and pages are specified in the Notes and Bibliography. The dates of the volumes are as follows: 1 (1960); 2 (1965); 3 (1971); 4 (1978); 5 — started 1979, but not yet complete. Supplement to volumes 1—3, started 1979, also not yet complete.

Transliteration of Arabic

The transliteration system of *The Encyclopaedia of Islam* has been followed, with three exceptions: 'j' not 'dj' is used for *jim*; 'q' not 'k' is used for *qaf*.

Consonants which are single in Arabic and double in Roman — e.g. 'kh' and 'sh' — are not underlined. Also, diacritical points under letters are given in authors' names in the Bibliography, but not in text or notes.

Acknowledgements

Figure 2.3, from Figure 1 in Norman A.F. Smith, 'Attitude to Roman Engineering and the Question of the Inverted Siphon', *History of Technology*, ed. A. Rupert Hall and Norman Smith vol. 1 (1976), courtesy of Dr Norman Smith; Figure 3.1, from Figure 2 in Norman A.F. Smith, *A History of Dams* (Peter Davies, London 1971), courtesy of Dr Norman Smith; Figure 6.2, from Figures 366 and 604 in *A History of Technology*, ed. Singer *et al.* (Oxford University Press, 1956), courtesy of Dr Trevor I. Williams; Figures 8.2, 8.3 and 8.8 are from Figures 14, 15 and 20 in J.G. Landels, *Engineering in the Ancient World* (Chatto and Windus, London, 1978), courtesy of Dr J.G. Landels; Figures 8.4, 8.5, 8.6 and 8.7 are from Figures 7, 54b, and 33 and 34 in Thorkild Schiøler, *Roman and Islamic Water-Lifting Wheels* (Odense University Press, 1971) — Dr Schiøler very kindly sent originals of these drawings and gave his permission to publish; Figures 8.9, 8.10 and 8.12 are from Figures 135, 138 and 142 in Al-Jazari, *A Compendium on the Theory and Practice of the Mechanical Arts*, ed. A.Y. Hassan (Institute for the History of Arabic Science, Aleppo, 1979), courtesy of Professor A.Y. Hassan; Figure 9.5 from Figure 2 in Stanley Freese, *Windmills and Millwrighting* (David and Charles, Newton Abbot, 1971), courtesy of David and Charles; Figures 10.2 from Figure 29 in Derek de Solla Price, *Gears from the Greeks* (Science History Publications, New York, 1975), courtesy of Professor Derek de Solla Price; Figure 10.4 (a), (b) and (c) from Figures 1a, 2 and 3 in W. Hartner, 'Asturlab' in vol. 1 of *The Encyclopaedia of Islam*, courtesy of the Editors of *The Encyclopaedia of Islam*; Figure 7.3 from *The Astrolabe* by Harold N. Saunders (Bude, Cornwall, 1971) — Mr Saunders kindly supplied the drawing and gave permission to publish. The remaining illustrations were drawn by the author.

All the Plates were supplied by Dr Norman Smith, who also kindly gave permission to publish.

Preface

This book describes the important engineering achievements of the peoples of Europe and western Asia in the period from 600 BC to AD 1450. A large area and a long time; some omissions were inevitable. These include military engineering, which requires a book to itself, and also a wide range of devices, textile machinery for example, that were largely manually operated. Even so, a process of selection and condensation had to be applied to every chapter in order (it is hoped) to leave the essentials intact. This is a technical history of engineering — social and economic factors are referred to only when they can throw light upon engineering developments. The History of Technology is a discipline in its own right, and should not be seen simply as an adjunct to other branches of history. Certainly, historians working in other fields need the findings of historians of technology, but they should be able to rely upon those findings. It is useless to construct a model of 'technology and society' in a given culture, if the technological data used are faulty. One aspect of this work is concerned with origins and diffusions, and another with *what* was being originated and diffused. Considerable attention is therefore paid to descriptions of techniques and machines.

The task of the historian of technology has been eased in recent years by the publication of texts and translations of source material, papers and monographs on particular topics and a few very good works of synthesis. My researches have included all these types of material but, clearly, not everything that is available. There comes a point, in preparing a work of this nature, when a halt must be called to further research, otherwise the book would never be written. I have tried to treat the area as a cultural whole, since I believe it to be so. I have therefore paid more attention than is usual to Islamic achievements, in an attempt to present a balanced view of engineering developments in the classical and medieval period.

I acknowledge with gratitude the generosity of The Royal Society in furnishing me with a grant to assist me in the preparation of this work. I wish to thank my friend Dr Norman Smith, of Imperial College, for providing me with illustrations and also for reading Chapters 3 and 4. His comments were very valuable, but the responsibility for any remaining errors or inconsistencies is, of course, mine. I am very gateful to Dr Thorkild Schiøler for sending me some of his own excellent drawings of water-raising machines, and for allowing me to publish them. Professor Derek de Solla Price, of Yale University, kindly allowed me to use an illustration of the Antikythera mechanism; for this, and for his advice and encouragement, I am sincerely grateful.

I also wish to thank my friend Professor Osker von Hinüber, of the University of Freiburg, for his advice on Indian culture and for providing me with books which would otherwise not have been available to me. I have found it a pleasure to work with the staff of Croom Helm, in particular Peter Sowden, and I thank them for their assistance, and for the quality of the production. I have found the library staff of the University of London and the Institute of Historical Research, both in Senate House, and of the Royal Asiatic Society consistently courteous and helpful. I am grateful to Dr John Watson of Great Bookham, who gave prompt and effective help when my health was jeopardised by pressure of work.

My dear wife Pat knows the depth of my feelings for her; this book could not have been written without her unfailing support and encouragement.

1　Introduction

The Background

A reference framework is required in order to locate events in time and space. With some contractions and omissions, Figure 1.1 shows the conventional divisions for the classical and medieval periods. Even before the birth of the idea of nationality, it is quite acceptable to refer to specific countries, such as Greece and Italy, whose boundaries are well defined. It is also usual to refer to areas in which there is felt to have been some degree of cultural unity — for example, the Roman Empire and Islam. Sometimes space and time are embraced by one image: the Roman Empire can mean either the first four centuries of our era or the area under Roman dominion. Used with care, these concepts have value for some historical purposes, but they can be very misleading. In the first place, we have to bear in mind the shifting of frontiers; in AD 750, for example, the Iberian peninsula was predominantly Muslim while Asia Minor was Christian — by 1450 the reverse was the case. Also, and this can be more serious, the conventional divisions are associated most closely with political and military realities, and often have little bearing on intellectual or social activities. When we think of Roman literature, the names of Tacitus, Virgil and Horace come to our minds, but in the richer, more populous East the languages were Greek, Syriac, Coptic and Aramaic. Most of the literary and scientific writings, in the eastern part of the Roman Empire, were in the first two of these languages. There can be no doubt, however, that the Roman period has distinctive features that justify its special place in history. The Romans were pre-eminent in the fields of organisation, administration, public works and domestic comfort. Their standards were not equalled in these respects until the nineteenth century.

According to convention, the Hellenistic Age began with the conquests of Alexander and ended with the death of Cleopatra. It therefore lasted for about three centuries — from about 330 BC to 30

1

Figure 1.1: Cultural Divisions

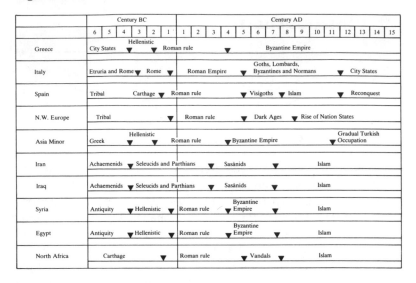

	Century BC						Century AD														
	6	5	4	3	2	1	1	2	3	4	5	6	7	8	9	10	11	12	13	14	15
Greece	City States		Hellenistic ▼ ▼		Roman rule				▼			Byzantine Empire									
Italy	Etruria and Rome ▼	Rome ▼			Roman Empire		▼	Goths, Lombards, Byzantines and Normans					▼	City States							
Spain	Tribal		Carthage ▼	Roman rule				▼ Visigoths	▼ Islam				▼ Reconquest								
N.W. Europe	Tribal			▼	Roman rule			▼	Dark Ages		▼ Rise of Nation States										
Asia Minor	Greek		Hellenistic ▼ ▼		Roman rule		▼ Byzantine Empire							Gradual Turkish ▼ Occupation							
Iran	Achaemenids ▼	Seleucids and Parthians			▼	Sasânids			▼		Islam										
Iraq	Achaemenids ▼	Seleucids and Parthians			▼	Sasânids			▼		Islam										
Syria	Antiquity	▼ Hellenistic		▼	Roman rule	▼	Byzantine Empire		▼		Islam										
Egypt	Antiquity	▼ Hellenistic		▼	Roman rule	▼	Byzantine Empire		▼		Islam										
North Africa	Carthage			▼	Roman rule			▼ Vandals	▼		Islam										

BC. During this period the Seleucids ruled in Syria, the Ptolemies in Egypt. It is often regarded as a period of stagnation, or even regression, between the glory of Greece and the grandeur of Rome. And this for an age that produced Archimedes, Theocritus and many major engineering works! There is, in fact, a strong case for prolonging the Hellenistic Age, for want of a better name, through the Roman Empire into Byzantine times and up to the advent of Islam. In the first centuries of our era great scientists such as Ptolemy, Pappos and Hero wrote in Greek in Alexandria, and there was also a thriving scientific tradition, with Syriac as its language, centred on Harran in northern Mesopotamia. Scholars from Harran were a seminal influence on the nascent science of Islam.

The concept embodied in the term 'Dark Ages', however, has some validity. The period from the fifth to the eighth century witnessed a marked decline in intellectual activity, and a falling off of standards in sanitation, water supply, communications and domestic comfort. The obscurity of the period is a serious drawback for historians, since the lack of written records means that for most subjects the only evidence comes from scanty archaeological findings. In the Byzantine and Iranian Empires, in this period, civilisation remained at a high level, but written evidence from these two cultures is very sparse. In 915 al-Mas'ūdī saw a large book dealing with the history of the Sasānid kings and many of the sciences. The book was found in the royal treasury in Persepolis in 731 and was translated from Farsi into Arabic for the Umayyad Caliph Hishām bin 'Abd al-Malik.[1] No copy of this work has been found, but the reference to it implies that there was intellectual activity in Sasānid Iran. Indeed, we have firmer evidence for the existence of a scientific tradition in Iran in the foundation of a famous hospital and medical school in the city of Gondeshapur in Khuzistan. The transformation of Gondeshapur into an important medical centre was due to a group of Nestorian Christians from eastern Anatolia who had fled from Byzantine persecution. This probably occurred in the reign of Khusraw I (AD 531–79). Although there may have been some Indian influence, the teaching and treatment, which was based solely upon scientific medicine, was derived from the schools of Alexandria and Antioch, but it became more specialised and efficient. Gondeshapur was to be the foundation of Islamic medicine in the eighth century.[2] There are more written records from Byzantium than from Iran in this period, but many of these deal with theological and philosophical matters. For neither area, in fact, — and this applies to Byzantium in the centuries after the rise of Islam — is there much data about technology.

The Arab conquests of the seventh and eighth centuries changed much, particularly in the close cohesion of religion and daily life that is characteristic of the Muslim religion. The success of Arabic in replacing the original languages throughout North Africa and most of the Middle East, and becoming the vehicle for literature in the whole of the Muslim world was of tremendous importance in the upsurge of intellectual activity that began when the process of conquest and consolidation was complete. A further impetus was the translation from Greek, sometimes through the medium of Syriac, of many scientific and philosophical works. The conquering armies from Arabia were, however, small in numbers compared to the populations of the conquered lands, and a slow process of fusion took place, with Islam and Arabic becoming predominant while the Arabs were influenced by the cultural traditions of the conquered peoples. When, in the present work, the word 'Islam' is used, it is to designate the cultural area that was the result of this fusion. It is necessary to bear this definition in mind, since many of the great writers in Islam have been non-Muslims. In some ways the term 'Arabic' is more satisfactory, but it leaves out of account, for example, the great literary works written in Farsi. Also, in discussing technological matters, we often have recourse to non-literary evidence. We can hardly call a mill built in Central Asia an 'Arabic' mill.

The standard of technology at any time depends mainly upon the demands of society. Where there are large urban communities to be fed, housed, clothed and provided with the raw and finished materials for commerce, we shall find technology applied to agriculture, communications and industry. The Arabic writers on geography in the tenth century describe a society in which the range of foodstuffs, the quality of textiles and indeed the standard of living in general were far in advance of conditions in Europe, and utilitarian technology was therefore relatively more developed. Courtly circles in Islam, not only in great centres such as Baghdad and Cordoba, but also in the courts of minor princelings, expected to be amused and given aesthetic pleasure by the production of ingenious devices, thus stimulating the development of fine technology, despite the apparent triviality of some of the devices. A further stimulus was provided by the requirements of astronomers for timepieces and for observational and computational instruments. (Astronomers, in fact, often made their own instruments.) Conditions were not propitious in Europe at this time. Populations were scattered in small communities and the ruling classes were largely illiterate. In the twelfth century European technology began to develop rapidly, along with the growth of cities and mercantile activity. In some fields, but not all, Europe had surpassed Islam by the close of the fifteenth century. The debt owed by Europe, in matters of technology, to Islam and other civilisations has never been fully acknowledged, and it is hoped that some of the material in this book will help to redress the balance. Nevertheless, it is undeniable that European technology has continued to develop from its medieval origins until the present day, while that of other cultures has not.

A close study of the interactions among technology, society and economic life is beyond the scope — and indeed the purpose — of this work. From time to time, however, connections will be indicated when these have a direct bearing on engineering developments. For example, the use of very large mills to grind corn for Baghdad is directly related to the fact that the population of the city, about 1.5 million in the tenth century, was a large net consumer of agricultural produce. It will also become evident that in certain places at certain times there were marked expansions in engineering activity, either in the scale of installations or in innovativeness, or both. We can identify the most significant of these:

Hellenistic world	330–30 BC. Machines, fine technology, irrigation.
Sasānid Iran	200 BC–AD 640. Civil engineering, irrigation.
Roman Empire	First four centuries AD. Civil engineering, surveying, water supply.
Islam	Ninth to thirteenth centuries AD. Water supply, irrigation, water power, fine technology.
Europe	Twelfth to fifteenth centuries (and beyond). Wind and water power, machines, mechanical clock.

Engineers and Artisans

The modern idea that the engineer is essentially an applied scientist simply does not hold water. It is true that in the last two centuries the engineer has had at his disposal a growing array of quantified scientific data upon which to base his designs, and that no engineer worthy of the name can ignore these data. Nevertheless, the execution of a construction project poses a number of problems that are unconnected with pure scientific reasoning. Some of these problems are not even of a directly technical nature: for example, the financial side of contracting and subcontracting with the accompanying measurement and valuation procedures. There are many other problems that are technical, but not directly related to science. An engineer needs to know the capabilities and limitations of men, machines and materials, and the restraints imposed by weather and soil conditions. Engineering science can assist in the resolution of some of these problems, but in the end we should expect our engineers to get their hands dirty — in the workshop or on construction sites. We should not be complacent, and assume that we have got the mixture of theory and practice exactly right. There have been enough failures in modern engineering to teach us a little humility. One might object, with some justice, that we approach the structural limits of materials much more closely than did the early engineers. So we should — with 4,000 years of their experience and almost two centuries of scientific analysis behind us!

In general, the classical and medieval engineers did not have a quantified, scientific basis for their designs. An exception to this statement is the case of the five simple machines — lever, wheel, pulley, wedge and screw. Mathematical analysis of these machines had begun to take shape among the Greeks of the fourth century BC. Their results were by no means wholly theoretical. It is worth quoting in full what Bertrand Gille has to say about the greatest early worker in this field:

> Archimedes of Syracuse (*c.* 287–212 BC), greatest of all the ancient exponents of mechanics and one of the greatest mathematicians of all time, was the first of the long line of those who have promoted science by propounding practical and definite problems. Like Pythagoras and Archytas, like the Egyptian and Babylonian surveyors before him, and like Leonardo and Galileo after him, Archimedes was a geometer because geometry is a technician's science. His researches on statics revealed the fundamental principles relating to the lever and the centre of gravity. His studies were of great assistance to those who sought to construct purposeful machines. These geometrical investigations of Archimedes were important, since, if we may believe Plutarch (*c.* AD 46–*c.* 125), they enabled him to calculate, for example, the number of pulleys needed for lifting a given weight with a given force. He knew how to calculate mechanical advantage, than which there is nothing more important for all lifting-devices.[3]

There can be no doubt that the results of Archimedes and his successors such as Hero of Alexandria (*fl.* middle of first century AD) were applied to lifting devices, war machines, steelyards and so on. The difficulty with so many other problems in engineering is that their solution almost always demands the use of differential or integral calculus, which was not invented until the seventeenth century. Nor should we forget the difficulties of making calculations without the place-number and decimal systems. We also have to bear in mind that engineers and master craftsmen were often illiterate, at least in the sense that they could not read the languages in which the theoretical treatises were written. This probably applies to most of the master masons of Europe, some of whom may have read their own vernaculars, but not Latin. But in the use made by the master masons of geometry lies the key to the way of thinking of all the early engineers:

> It becomes evident that the 'art of geometry' for mediaeval masons meant the ability to perceive design and building problems in

terms of a few basic geometrical figures which could be manipulated through a series of carefully prescribed steps to produce the
points, lines and curves needed for the solution of the problems.
Since these problems ranged across the entire spectrum of the
work of masons — stereotomy, statics, proportion, architectural
design and drawing — the search by modern scholars for the
geometrical canons of mediaeval architecture is appropriate
enough, so long as we keep clearly in mind the kind of geometry
that was actually used by the masons. The nature of that geometry
suggests that these canons, when recovered, will not be universal
laws which will at last provide *the key* to mediaeval architecture;
rather, they will be particular procedures used by particular master
masons at particular times and places.[4]

It is precisely this kind of constructive approach that determined
how a given problem was to be solved. The engineers had to know
geometry, arithmetic and some trigonometry. They also had to
understand the properties of materials and the way in which fluids
and solids behaved in a certain set of conditions. Faced with a
problem, they applied this knowledge to solve it in the most satisfactory manner open to them. For example, in bridge building it is
desirable to keep the number of piers to a minimum. Piers are an
obstruction to shipping, they are expensive to construct and they are
more prone to failure than the superstructure. If semicircular arches
are used for a large span they will be very high, so it became necessary
to build segmental arches. To take another example, from mechanical engineering, the suction pump with horizontal cylinders and
vertical suction pipes was developed by Islamic engineers from the
so-called 'Byzantine siphon', used to discharge Greek fire from
warships. In the earlier force-pumps the vertical cylinder had stood
directly in the water, but clearly this was both undesirable and
impractical with inflammable fluids. Both these cases, and many
more could be cited, illustrate the solution of problems by practical
thinking — the theory came later.[5]

The concept of specialisation is very recent. Of all the scholars
whose names have come down to us, only al-Jazarī (late twelfth
century) seems to have devoted his entire life to engineering. The
others began their careers in a variety of occupations: Ctesibius (third
century BC), according to Vitruvius, was a barber; Guido da
Vigevano (first half of fourteenth century AD) was trained in
medicine; Mariano Taccola (first half of fifteenth century AD) was an

artist and sculptor. Al-Bīrūnī (d. after AD 1050), probably the greatest scientist of medieval Islam, made astronomical instruments and studied mining technology. Although some of the wide-ranging studies of men such as these may have been decided by personal preferences, they were often obliged to follow the dictates of their patrons. No doubt there was an element of prestige in having learned men attached to one's court, but they were expected to earn their keep as physicians, astronomers, teachers, architects and engineers. Demarcation was unknown — they had to apply themselves to whatever matters were the concerns of their patrons at any given time. Versatility was normal, but it was also essential. It is hardly surprising to find that many of the engineers were concerned with devising and building engines of war — Archimedes, Vitruvius (first century BC), Hero, the Banū Mūsà (mid-ninth century AD), de Vigevano and Taccola — to name but a few. Many of the English medieval architects also built siege engines.[6]

Considerable discussion has taken place on the status of engineers in society at various times. (Here we are defining engineers as those who devoted at least part of their time to engineering.) It is well known that Plato and other classical writers had a low opinion of technology and manual work, but these activities were not so universally despised as might appear from these writers.[7] Indeed, it seems to have been quite usual for engineers to occupy a respected place in society. Frontinus (first century BC) was a highly-placed government official, the Banū Mūsà were the trusted advisers of a succession of Abbasid Caliphs in Baghdad, Taccola became an eminent citizen of his native Siena. The great architects of the Middle Ages, often appointed directly by the Crown, undertook the design, administration and supervision of construction on large public buildings. It is impossible, however, to make any general statement about the status of engineers, partly because we are considering a number of widely differing societies over a long span of time, and partly because the background, training and duties of engineers varied so greatly. We know that some of them, such as al-Jazarī and the English architects, were themselves master craftsmen who were highly skilled in the various engineering trades. Others, such as Frontinus, were simply civil servants who happened to be placed in charge of departments that were concerned with engineering matters. Between these two extremes lies the whole range of classical and medieval engineering. In dealing with the first category we can see quite clearly the difference between engineers and artisans — the engineer was simply an artisan who had reached the top of his profession. It was often possible, at certain periods, for a master mason, carpenter or metalworker to reach a position of high responsibility.

It is of course obvious that there have always been large numbers of workers — skilled and unskilled — who never have risen high in the ranks of society. The question of the varying status of artisans throughout the classical and medieval periods is a topic of great complexity, which it is beyond the scope of this book to discuss in any detail. At certain times artisans, even if they were not actually slaves, were subject to strict control. In Imperial Rome the *fabricenses*, or armament workers, were branded on the arm so that if they deserted the factories they could be identified and returned.[8] When the Caliph's wazir, or counsellor, was constructing a bridge in Iran in the second half of the tenth century he conscripted the engineers and craftsmen from Isfahan.[9] On the other hand, the work of craftsmen was often well respected; in Iran the names of carpenters sometimes appeared as inscriptions on their products, for example the name of a certain Husayn b. 'Alī on a richly carved door made in 1428.[10] But for any kind of detailed study of the question, recourse must be had to works of economic and social history.[11] There is also one type of writing that is particularly useful for understanding the work of craftsmen in Islam. These are the books written for the instruction and guidance of the *muhtasib*, who was the official appointed by the ruler or his representative to supervise the affairs of the market. The duties of this official were wide ranging, since the market, or *sūq*, was the centre for commerce and industry in the city, and he was required to supervise its daily affairs: the maintenance of moral standards and religious observance; quality and quantity control over retailers and manufacturers; sanitation and water supply; checking the manufacture of building materials and the erection of houses. Usually an experienced craftsman was made responsible to the muhtasib for the maintenance of standards in his own trade. There is therefore a great deal of useful information in the *hisba* manuals about manufacturing and constructional methods.[12] Nevertheless, many of the trades practised in the Islamic, and indeed European, cities were not concerned with engineering and, in any case, we cannot readily define the relationships that existed between the engineers, and the craftsmen of the cities.

Nowadays there is a marked distinction between artisans who are based in workshops and factories and those who usually work on construction sites, although there is of course mobility from one type of work to the other. For many trades, however, the type of skills demanded on construction are quite different from those needed in a workshop and this was equally true in earlier times. The type of woodworker who made ornamented furniture would not have been able to erect the timber falsework for a large masonry arch. The available evidence does indeed indicate that engineering workers formed a distinct group in classical and medieval times. Frontinus tells us, for example, that there were two gangs of slaves allocated to the maintenance of Roman water installations, one numbering about 240 men, the other 460. Both were divided into several classes of workmen: overseers, reservoir-keepers, inspectors, pavers, plasterers and other workmen.[13] In medieval England masons were often conscripted for Royal works, whereas for non-Royal buildings they were hired by the master mason.[14] In either case, it is probable that the same men were employed on a single site for a period of some years. Some classes of work did not demand large numbers of skilled artisans. For the excavation of canals in eleventh-century Iraq, for example, the engineers and surveyors established the lines, levels and profiles of the canals, and gangs of labourers worked under their supervision.[15]

Things are rather different in the field of fine technology. Al-Jazarī, who composed his excellent machine book in AD 1206, was clearly a master craftsman in his own right, who was capable of constructing large and small machines entirely with his own hands.[16] This consisted of metalwork of all kinds, including casting in copper, brass and bronze, soldering and tinning, sheet metalwork and so on. For his larger devices he would have needed the help of a labourer, and he may have trained apprentices, but otherwise he required no outside assistance. The treatise of Theophilus, written in the early 1120s, is the work of a master craftsman in paint, glassmaking and metalwork, although the engineering content of his work is much less than that of al-Jazarī.[17] The case of the Banū Mūsà, who composed their book on Ingenious Devices in Baghdad about 750, is rather different. These three brothers were prominent engineers and scientists, and their work shows a masterly use of differential pressures, but they make very little mention of the processes of construction, although the devices 'as built' are described clearly enough. It is possible, though by no means certain, that they did not make the devices themselves, but employed skilled metalworkers who worked under their direction.[18]

A Note on Tools

Man has been making tools for at least 2.5 million years and the stone tools in their last phase of development, the Neolithic, include most of the main types in use today. The advent of iron, however, was a very significant step in the history of tools, particularly with the discovery of hardening by carburisation and heat treatment, which led to superior edge tools of great toughness. These processes were discovered early and were fully utilised by Roman times. The basic kit of iron tools was available to Assyrian craftsmen in the eighth century BC and remained in use through classical and medieval times — the only major addition was the plane. But tools, within a given tool type, became very specialised. For percussion tools, for example, there was a complete range, from the heavy hammers used in quarrying and forging to the delicate hammers of the goldsmiths and jewellers.

The basic cutting tools were the axe and the adze — in the former the blade is parallel to the handle, in the latter it is at right angles to the handle. The heavy felling axe was used for clearing woodland, whereas the broadaxe was designed to convert felled timber to squared. It had a shorter handle than the felling axe with a heavier head, and was a two-handed tool. The adze was an indispensable tool of general utility for, in addition to surfacing, it was particularly useful for trueing and otherwise levelling framework such as posts, beams and rafters. Saws were originally designed to cut on the pull stroke but the modern practice of using the push stroke probably came into use towards the end of the Roman period as saws became more robust. A major contribution to saw design was noted in the first century AD by Pliny the Elder; the teeth of the saw were set — that is, alternate teeth were bent to one side or the other, so that a slot was created in the wood wider than the thickness of the blade. This helped to discharge the sawdust and also allowed the saw to run with less friction.

Chisels and gouges were made in great variety, from the heavy tools of the carpenters and masons, which were struck by mallets, to the more delicate hand-held tools of the woodcarvers. The Romans were the first definitely known users of the plane, the earliest examples coming from Pompeii. They appear to be without a pre-history, without even vague antecedents. The modern plane differs in detail, but not in principle nor in general appearance. The essential features of the plane are the wooden body pierced by a wide throat with a narrow slit in the bottom. The blade is held in position by a wedge tapped under an iron bar across the throat. The file is also used for smoothing, and is more suitable for metalworking than the plane. The file's many tiny chisel-like teeth were produced by cutting with a small chisel-like hammer or with a hammer and chisel. After this treatment the file is tempered and quenched. Theophilus mentions files of square, round, triangular and other shapes. Specialised uses of the file included sharpening saws and cutting the teeth of gearwheels.

Drilling and boring was done with simple awls and gimlets that push the material aside without removing it, with drills and with augers. The bow drill was invented as early as Upper Paleolithic times and it remains in use today in some parts of the world. A bowstring wrapped around the shaft imparts rotation to the shaft in alternating directions. The more complicated pump drill was developed in Roman times. A crosspiece that could slide up and down the spindle was attached by cords that wound and unwound about it. Thus a downward push on the crosspiece imparted a rotation to the spindle. A flywheel on the spindle kept the motion going, so that the cords rewound in reverse to raise the crosspiece as the drill slowed, and the next push down brought the spindle into rotation in the opposite direction. The basic auger originated in the Iron Age. It resembled a half pipe sharpened on the inside, along the edge, or both. It had a crossbar that could be turned with two hands. The earliest illustration of the brace-and-bit occurs in an illustration of about AD 1425.

The only early machine tool was the lathe, which was introduced into Europe by the Greeks in the seventh century BC, probably from the wood-producing areas of the Middle East. Most early machines were pole-lathes. This consisted of a bed with uprights in which bearings were installed; the work, roughly hewn to a cylindrical shape, was revolved in these bearings. A resilient pole projected over the bed, and a cord attached to its end was wrapped round the work and thence attached to a hinged board serving as a treadle. When the board was depressed the cord spun the work round in one direction; when the pressure of the foot was relaxed the bough sprang back, drawing cord and treadle upwards and turning the work in the opposite direction. Motion was thus reciprocating, not continuously rotary, and this allowed the wood cut by the simple hook-tools on the down-stroke to be cleared on the up-stroke, thus avoiding clogging and loss of control.

Other tools, in use throughout our period, include tongs, pincers and pliers, engraving tools, and punches for doing repoussé work in sheet metal. Measuring and defining tools included the plumb line, level, square and compasses. Some applications of these will be described when methods of construction and assembly are dealt with in subsequent chapters.

Notes

1. Abu'l-Hasan al-Mas'ūdī, *Kitāb al-tanbīh wa'l-ishrāf*, ed. M.J. de Goeje, vol. 3 of *Biblioteca Geographorum Arabicorum* (Brill, Leiden, 1894), pp. 106-8.

2. Aydin Sayili, 'Gondēshāpūr', *EI*, vol. 2, p. 1120.

3. Bertrand Gille, 'Machines' in Charles Singer, E.J. Holmyard, A.R. Hall and Trevor I. Williams (eds.) *A History of Technology* (Oxford University Press, 1956, reprinted 1979), vol. 2, pp. 629-57, pp. 632-3.

4. Lon R. Shelby, 'The Geometrical Knowledge of Mediaeval Master Masons', *Speculum*, vol. 47 (1972), pp. 395-421, pp. 420-1.

5. For the segmental arch see Chapter 4, for pumps see Chapter 8. The line of reasoning in this paragraph was generated in the course of discussions between myself and Dr Norman Smith of Imperial College, London.

6. John Harvey, *English Medieval Architects* (Batsford, London, 1954), pp. 22ff., 91, 150ff.

7. F. Klemm, *A History of Western Technology*, tr. Dorothy W. Singer (George Allen & Unwin, London, 1959), pp. 18-24.

8. W.H.G. Armytage, *A Social History of Engineering*, 3rd edn (Faber and Faber, London, 1970), p. 33.

9. Zakariya bin Muhammad al-Qazwīnī, *Athār al-bilād* (Beirut, 1960), p. 303.

10. Hans E. Wulff, *The Traditional Crafts of Persia* (The MIT Press, Cambridge, Mass., 1966, 2nd printing 1976), p. 81.

11. Useful discussions of the subject will be found in N.J.G. Pounds, *An Economic History of Medieval Europe* (Longman, London, 1978), pp. 27-34, 71-9, and especially Chapter 4, 'Medieval Manufacturing', pp. 279-337.

12. One of the best (and most accessible) of the *hisba* manuals is Ibn al-Ukhuwwa. *Ma'alim al-Qurba fi Ahkām al-Hisba*, Reuben Levy (ed.) with Arabic text, notes and abridged English translation (Gibb Memorial Series, New Series, London, 1938).

13. Frontinus, *The Strategems and the Aqueducts of Rome*, ed. wth Latin text and English translations by Charles E. Bennett and Mary B. McElwain (Loeb Classics, Cambridge, Mass. and London, 1925, reprinted 1980), pp. 447-8.

14. Lon R. Shelby, 'The Role of the Master Mason in Mediaeval English Building's, *Speculum*, vol. 39 (July 1964), no. 3, pp. 387-403, p. 396.

15. Claude Cahen, 'Le Service de l'irrigation en Iraq au début du XIe siècle', *Bulletin d'etudes orientales*, vol. 13 (1949-51) pp. 117-43.

16. Ibn al-Razzāz al-Jazarī, *The Book of Knowledge of Ingenious Mechanical Devices*, tr. and annotated by Donald R. Hill (Reidel, Dordrecht, 1974).

17. Theophilus, *On Divers Arts*, tr. from the Latin with introduction and notes by John G. Hawthorne and Cyril Stanley Smith (Dover Publications Inc., New York, 1979).

18. The Banū Mūsà, *The Book of Ingenious Devices*, tr. and annotated by Donald R. Hill (Reidel, Dordrecht, 1979).

19. There is no comprehensive work on tools. This note is taken from the following: Richard S. Hartenberg, 'Hand Tools', in *Encyclopaedia Britannica*, 15th edn (1974) pp. 605-24 (with bibliography); R.W. Symonds 'Furniture: Post-Roman', pp. 221-58, and R.H.G. Thomson, 'The Medieval Artisan', pp. 383-96,

both in *A History of Technology*, ed. Singer *et al.*, vol. 2 (Oxford, 1956 reprinted 1979); the section on metalworking tools in Theophilus, pp. 81-95 is of the greatest importance.

PART ONE:
CIVIL ENGINEERING

2 Irrigation and Water Supply

History

We are concerned here with the study of the engineering elements in the two related technologies of irrigation and water supply. Indeed, so closely are they related that several of their constituents are identical. This applies to the main arteries bringing water to the distribution networks — canals, aqueducts and qanats — and to the methods of impounding and storing the water. As we shall see, the same arteries were sometimes used to serve both purposes. Irrigation and water supply stimulated the development of other technologies, which are dealt with elsewhere in this book: Dams in Chapter 3; Bridges (including aqueducts) in Chapter 4; Surveying in Chapter 7; and Water-raising machines in Chapter 8. Some applications of these technologies will, however, be mentioned in this chapter.

From an engineering point of view irrigation preceded water supply. The ancient civilisations based upon irrigation in Mesopotamia and Egypt had been in existence for over two millenia before the start of our period and although they had known times of decline, there can be no doubt that irrigation in the classical, Hellenistic, Sasānid and Islamic cultures was based upon these earlier systems. Apart from the introduction of new types of water-raising machines (by no means a negligible development) it cannot be said that any radically new techniques were invented to add to the repertoire of the Egyptian and Mesopotamian irrigators. It could scarcely be otherwise: the basic problem of impounding the water, conducting it to the land and finally draining the surplus remains as it has always been. To judge the engineering of any period or culture, however, solely by the number of new inventions produced is to fundamentally misunderstand the nature of engineering. From ancient times onwards 'the development of numerous facets of civil engineering was of fundamental importance. The construction of dams and canals, matters to do with water flow and control, and elaborate surveying problems, all

17

presented themselves uncompromisingly. Their successful solution was in the hands of experts.'[1] Not only the construction, but also the maintenance of the systems demanded the constant exercise of engineering and administrative skills.

Before we examine the history of irrigation in classical and medieval times we need to discuss briefly the four different types of irrigation. *Basin* irrigation consists of levelling large plots of land adjacent to a river or a canal, each plot being surrounded by dykes. When the water in the river reached a certain level the dykes were breached, allowing the water to inundate the plots. It remained there until the fertile sediment had settled, whereupon the surplus was drained back into the watercourse. The regime of the Nile, with the predictable arrival of the flood, makes Egypt particularly suitable for basin irrigation, but it was also practised elsewhere. A Sumerian text dated to about 1700 BC describes clearly the use of the technique in Mesopotamia.[2] In this case the crop was barley and the field was inundated five times, once before the seed was sown and four times while the barley was growing. *Perennial* irrigation was, and is, practised extensively in the Mesopotamian plain. As the name implies, it consists of watering crops regularly throughout the growing season by leading the water into small channels which form a matrix over the field. Indeed, a network of waterways is typical of perennial irrigation. Water from the main artery — a river or a major canal — is diverted into supply canals, then into smaller irrigation canals, and so on to the fields. In many cases the systems operate entirely by gravity-flow, but water-raising machines are used to overcome obstacles such as high banks, natural or artificial. Perennial irrigation from wells, using devices of varying degrees of sophistication to raise the water, has also been practised from very early times.

Terrace irrigation was used at an early date in Syria and Palestine, and in India, China and pre-Columbian America. The last locale is important, since it indicates that the technique did not have to diffuse from a single point of origin. Terrace irrigation is a method used in hilly country and consists of the formation of a series of terraces stepped down a hillside. The effort required is high in relation to the levels of production, but if the land is the sole livelihood of a family or a community there is a no alternative. It is impressive and rather moving to see the terraced hillsides in the mountains of Lebanon or Italy, for example, where some of the plots are only a few square metres in area. Irrigation is by means of stored rainfall, wells, springs and occasionally qanats.

Wadi irrigation, the last category, is in some ways the most impressive, since it depends, not upon the flow of great rivers, but upon sporadic rainstorms in otherwise arid lands. In the Yemen agriculture was sustained by the occasional heavy rainfalls which fall in the west of the country and feed water to the Wadi Dhana. Across this watercourse, about three miles above the capital Marib, one of the largest dams of ancient times was built. Following its original construction in the eighth century BC it was successively raised, not to impound water for long periods but to raise the wadi floods to increasingly higher levels in order to irrigate more and more land by means of a canal system which used the wadi itself as a drain. The breaking of the dam at Marib, symbolising the breakdown of the irrigation system of the Yemen, is mentioned in the Koran.[3] Inscriptions have been found recording the restoration of the dam about the middle of the fifth century AD and again at the middle of the sixth; the final breakdown is thought to have been in the quarter-century before Muhammad's birth.[4] From the second century BC until about the beginning of the first century AD the Nabataeans produced a flourishing civilisation in southern Palestine and Jordan. Using wadi irrigation they developed a thriving agriculture in the Negev that was unequalled until modern times.[5] Whereas irrigation in the Yemen depended upon a single large dam, the Nabataeans built thousands of little barriers sited across one wadi after another in order to divert or capture the one or two weeks of runoff occurring each year. The system included elements of terrace techniques because the passage of silt-laden water gradually filled the reservoirs and stepped off each wadi into a terrace-like system.

We probably know more about the irrigation systems in ancient Egypt and Mesopotamia than we do about the systems in those countries in Hellenistic and Sasānid times. Even for Islam the study of irrigation has been patchy, and has tended to concentrate upon Egypt, Iraq and the Iberian peninsula. One reason for our lack of information about Hellenistic and Sasānid times is the scarcity of documentation, although for the former the existence of thousands of papyri recording matters of taxation and land usage has materially increased our knowledge. It does seem that Egypt under the Ptolemies witnessed a marked increase in agricultural activity and an improvement in techniques. The introduction of iron tools and of new types of water-raising machines (see Chapter 8) contributed to this increase, which may well have been stimulated by the demands upon agricultural produce of the great city of Alexandria. In Iraq major shifts in the course of the Tigris and the neglect of the irrigation systems in the later Middle Ages have rendered the assessment of earlier irrigation systems very difficult, particularly when we realise that the great expansion of irrigation in Muslim times has, by its very success, obliterated most of the evidence of earlier systems. It is apparent, however, that Sasānid engineers had made considerable improvements to the irrigation network in Iraq, including extensions to the great Nahrawan canal, to the east of the site of Baghdad.[6] The capital of the Sasānid Empire was Ctesiphon (al-Madā'in to the Arabs), a large city that was certainly a net importer of agricultural produce. It was located on the Tigris, about 20 miles to the south of modern Baghdad.

The Romans inherited the irrigation systems in the lands which they occupied. In Italy itself irrigation was never important, but it is probable that it was practised in the coastal areas of the Iberian peninsula, especially in the regions that had been colonised by the Carthaginians and the Greeks. The extent of Roman irrigation in Iberia, however, and of their Visigothic successors, has not yet been thoroughly studied. The lack of a firm datum seriously undermines efforts to understand later developments, especially those of Muslim origin. Egypt was the granary of Rome and we can therefore be sure that the irrigation systems founded in Pharaonic times and further developed in the Hellenistic Age were fully utilised. After the conquest of Palestine by Trajan in AD 106 the Nabataean ideas of wadi irrigation were rapidly assimilated and applied in North Africa, where they formed the basis for an expansion in agriculture along the North African coast, which sustained, among others, the magnificent cities of Leptis Magna and Sabratha. The latter has the same effect upon the traveller as the ruins of Palmyra; both come into view quite suddenly, splendid ruins now surrounded by wastelands.

Thomas Glick, in one of his penetrating studies of Muslim influences on Spanish agriculture and society, has summarised the diffusion of irrigation techniques as follows:[7]

1. the locus of invention is the ancient Near East, particularly Persia; then
2. a stage of limited diffusion through the Mediterranean in classical times, most typically under the Roman aegis; next,
3. an intensification of use and perfection of the technique by the Muslims, eighth to tenth centuries;

4. a particularly dramatic intensity of use in Al-Andalus, which in turn becomes

5. a new centre of diffusion back to North Africa and, later, to the New World.

In the passage quoted, Glick is introducing a discussion of four techniques, taken as examples that evaluate the role of the Arabs, and later the Spaniards, in the diffusion of irrigation technology. These are (a) diversion dams; (b) the noria; (c) qanats (see below), and (d) the cultivation of oranges. A reservation must be made to his otherwise admirable summary, in his identification of the main locus of early invention in Persia. The qanat certainly originated in Persia or Armenia, but the noria and the use of diversion dams, together with other irrigation techniques probably came from other regions. With this reservation, Glick's statement provides us with a sound summary of the diffusion of techniques in western Asia and the Mediterranean. As mentioned earlier, we do not know the extent of the irrigation practised in the Iberian peninsula under the Romans and their Visigothic successors, but we can be sure that a great expansion, both in the area irrigated and in technology, took place in the peninsula after the Muslims were established. One confirmation for this statement comes from the fact that the Muslims introduced a number of new crops into Iberia. Some of these, such as rice and sugar-cane could only be grown under irrigation, while others were temperate species that could only be stabilised in a semi-arid environment by irrigation.[8]
When the Muslims conquered Spain in AD 711, the Arab Empire was ruled by the Umayyad dynasty from Damascus. In 750 the Umayyads were overthrown in the East by the Abbasids, who moved the capital to Baghdad, but the Umayyads held on to the Iberian peninsula, where their Emirs (later Caliphs) ruled from Cordoba. This is one of the factors which accounts for the 'Syrianisation' of the Spanish landscape, another being the similarity of the climate and hydraulic conditions in parts of Spain to the conditions obtaining in the Ghūta — the large oasis surrounding Damascus, watered by the Barada and other rivers.[9] It is likely that the Syrian pattern was imposed upon the cultivators, mainly Berbers, by the Umayyad Emirs in the first quarter of the ninth century. In any case, there were few large rivers in the Berbers' North African homelands, and they were therefore willing to apply the more appropriate eastern methods in Spain, where river irrigation was possible on a much larger scale. In

places where the rivers supplied sufficient water, for example the regions of Valencia, Gandia and Murcia, the methods of irrigation and its administration were based firmly upon Syrian models. Further south, in the oasis-like communities of Elche, Novelda and Alicante, where the water sources were typically springs rather than rivers, irrigation water was also distributed by canals, but the administrative arrangements were different from those in the northern river valleys. In the latter, water rights were controlled by the State, and water was distributed on a proportional basis. In the former, water rights tended to be in private hands, alienable from land, and allocation was by fixed time units which could be bought and sold.[10] As the Christians gradually reconquered the Iberian peninsula, the Muslim irrigation systems were taken over more or less intact.[11] Many words of Arabic origin referring to irrigation technology entered the Spanish vocabulary, and are an eloquent testimony to the Muslim origin of many aspects of Spanish agriculture and irrigation.

Because Muslim irrigation in Spain (together with other technologies) had a direct bearing on developments in Europe, it has received the attention of western scholars such as Thomas Glick. We must give our wholehearted applause to studies of this kind, but should also bear in mind the great undertakings in the construction and maintenance of irrigation systems that took place in the lands of the eastern Caliphate from the seventh century onwards. These developments reached their climax in the tenth century, after the power of the Abbasid Caliphs had been dissipated, but while the Islamic world, as a cultural entity, stretched from the Atlantic to Central Asia. The great Arabic writers of the tenth century were able to move freely (though not always safely) over this vast area, and have left us lively and informative accounts of what they saw in their travels. Because they were interested in all aspects of society, we need to study their works thoroughly in order to arrive at an understanding of any given topic, but the effort is worthwhile in view of the information that can be gathered about so many subjects. There is a large amount of information about irrigation and water supply, and indeed it would not be difficult to write a book dealing solely with Muslim irrigation in Iraq. The story is complicated by major shifts in the courses of the Tigris and Euphrates, and by variations in the area of swampland and lagoons which lies to the north-west of Basra.

In general, the Sasānid kings had devoted considerable attention to the construction and maintenance of canals for irrigation and drainage, and to keeping the dykes in repair. This was a difficult task,

Figure 2.1: Hydraulic Map of Medieval Iraq

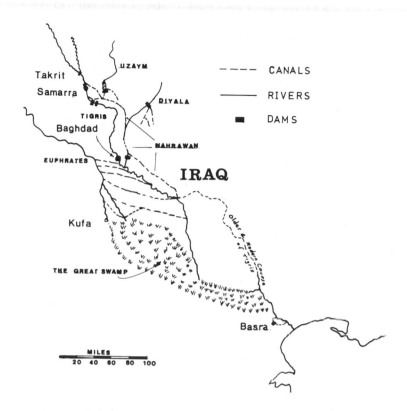

since the plain below Baghdad is flat and the two rivers are subject to disastrous flooding. One such occurred about the year AD 629, resulting in major changes in the courses of the rivers, and in a great extension to the swamplands. Figure 2.1 shows the topography in Iraq in medieval times. The Great Swamp then covered an area of about 200 miles by 50. At its northern end the Euphrates discharged into the swamp through its main course, which flowed past the city of Kūfa; its present course, flowing past Hilla, was then a great irrigation canal called the Nahr Sūra. The Tigris flowed well to the west of its present course, which was also its course before the floods of the seventh century. It flowed past the city of Wāsit (both Wāsit and Kūfa, once great cities, have now disappeared) and entered the Great Swamp at Qatr. Both rivers emerged from the swamp a few miles above Qurna, where they combined, as they do today, to form the waterway known as the Shatt al-'Arab, then called the Blind Tigris. Along the northern edge of the Great Swamp, from Qatr to a point upstream from Qurna, a line of lagoons linked by open channels made navigation possible from Baghdad to Basra. It is not known precisely when the Tigris shifted back to its older bed and when the Euphrates took its present course, but there seems to have been a gradual change, starting about 1200 and being completed in the sixteenth century.[12]

The Muslims inherited and developed the Sasānid irrigation system, the major expansion occurring after the founding of Baghdad in AD 762. The basic Sasānid network was dictated by topography, since there is a slight eastward tilt in central Iraq, and hence the gridiron of large canals ran from the Euphrates into the Tigris. The main artery for the irrigation of the lands to the east of the Tigris was the great Nahrawān canal, which left the Tigris a short distance below Takrīt and rejoined the river about a hundred miles below Baghdad. The upper part of the canal was known in Arabic as Al-Qātul al-Kisrāwī — 'the Cut of the Chosroes' — since this part of it had been excavated under the Sasānid kings. The rivers 'Uzaym and Diyāla discharged into the Nahrawān from the east, and these rivers were dammed by Abbasid engineers to provide water for a huge irrigated area. Important canals taking off from the west of the Nahrawān system included the Khālis and the Bīn; the waters of these made possible a closely-cultivated area north of Baghdad, and in part supplied the city itself.[13]

In the south of Iraq, due to the influence of the tides on the Shatt al-'Arab and the lower reaches of the Tigris and Euphrates, there is a particularly favourable situation for irrigation. Thanks to a system of canals specially adapted to the situation, there is no need to use machines to raise the water to the fields, since there is always sufficient raised by the tides to irrigate the fields. The cleansing of the canals is done automatically by the movement of the tides, and drainage is effected during the ebb tide. There is therefore no danger of salination, despite the presence of dissolved salts in the water due to the proximity of the sea.[14] The Basra area presents a marked contrast to central Iraq in the summer months, since the vast plantations of date palms provide shade for flourishing vegetable gardens, whereas further north the dust is pervasive and the canals are half empty, the water remaining being brackish.

Al-Balādhurī (d. *c.* 892) has left us an account of the foundation of Basra and the measures subsequently taken to supply the city with water for irrigation and drinking.[15] When 'Utba bin Ghazwān was the commander of a Muslim army in southern Iraq in 638 he selected the site of Basra, after consultation with the Caliph 'Umar I, as an encampment for his troops. It was located about 15 km to the west of the Shatt al-'Arab, and in the early years it was simply an army camp consisting of huts made from reeds, which were dismantled when the troops were absent on campaigns. In the early days drinking water had to be transported from the Shatt al-'Arab, and although attempts seem to have been made to excavate canals to the site from the river in the Caliphate of 'Umar I, neither of the two main canals was completed until after 660, when the Umayyads assumed power. One of these great canals, the Nahr Ma'qil, came down from the north-east and carried shipping from Baghdad; the other, the Nahr al-Ubulla, carried ships going south-east to the Gulf. The canals were linked by another canal, upon which the city itself was located. A great many irrigation canals were then excavated, and Basra became a thriving centre for agriculture, and the most important city in Iraq for commerce, finance and learning. Although it was eventually eclipsed by the rise of Baghdad it was still prosperous in the tenth century; al-Istakhrī described its enormous network of canals and its abundant agricultural produce.[16]

Although the cases of Iraq and Egypt come naturally to mind when we think of Muslim irrigation, almost every province of the Islamic world used the technique to a greater or lesser extent. We have descriptions of irrigation systems and the crops that they watered from Spain and North Africa, Egypt, Iraq, Iran and Central Asia. Only in Syria and Palestine does most of the cultivation seem to have been by rainfall. In medieval times the land between the upper reaches of the Tigris and Euphrates was known as al-Jazira; today it lies partly in Iraq and partly in Turkey. Al-Jazira was a supplier of corn and other foodstuffs to Iraq, and agriculture was partly by rainfall and partly by irrigation. One of the most fertile provinces of Islam was Khurasan, then very much larger than the modern Iranian province of that name, since it included parts of present-day Afghanistan and the USSR. It included a number of large cities and exported agricultural produce to other parts of the Islamic world. One of its most fertile areas was centred on the city of Marv, on the river Murghab, which provided the irrigation water for extensive farmlands. In the tenth century the superintendent of irrigation at Marv is said to have had more power than the prefect of the city, and to have supervised a workforce of 10,000 men.[17]

Surpassing all these, however, was the province of Sughd — ancient Sogdiana, now part of the Uzbek SSR. The mainstay of its fertility was the Sughd river, now called the Zarafshan, which flowed through the great cities of Samarkand and Bukhara. Also important was the smaller river to the south upon which the cities of Kish and Nasaf were located. Both rivers ended in marshes or shallow lakes in the western desert towards Khwarizm. Sughd had attained its greatest splendour in the ninth century, under the Iranian dynasty of Samanid Emirs, but in the following century it was still rich and fertile beyond compare. In the tenth century Bukhara consisted of a central walled city measured a league across in every direction, and the city was surrounded by towns, palaces and gardens, which were in turn encompassed by a wall that must have been over 100 miles in circumference. The Sughd river, with its numerous canals, passed through this great enclosure. About 150 miles upstream lay the city of Samarkand, situated a short distance from the southern bank of the river, and occupying high ground. For many miles around the city the fertile lands extended, watered by innumerable canals. The produce of Sughd included corn, rice, fruits and nuts of all kinds, and timber. The Arabic writers of the tenth century all devote many pages to describing the richness of the province. They also describe the canal systems in some detail, but unfortunately it is not possible to make an accurate reconstruction of these systems from the data which they supply. The cities of Bukhara and Samarkand, and indeed the whole of Transoxiana, were destroyed by the Mongols in 1219, but regained much of their former glory when Timur made Samarkand his capital at the close of the fourteenth century.[18]

Water supply, as an engineering enterprise, is concerned mainly with urban communities. Since man needs at least one and a half pints of water a day simply to stay alive, efforts to obtain an adequate supply have always been part of human existence. Good drinking water occurs naturally in rivers, lakes, springs, oases and underground, and rainfall may also be collected in natural or artificial containers. The incidence of springs in arid or desert lands is uncertain, yet even today primitive societies have an uncanny ability to locate sources of water and to plot their courses from one source to another. The tapping of underground sources by sinking walls is even more of a mystery. It has been conjectured that the practice arose in conjunction with the driving of shafts in the search for metallic ores, but it is just as likely that men dug for water where they needed it, and were frequently disappointed. In any event, the location of good, reliable sources of water was an important factor in the location of the earliest settled communities.[19]

Irrigation preceded water supply as a technology because the large cities of the ancient world were located in river valleys where the supply of water was assured. Even so, although the waters of the Nile, Tigris and Euphrates may not have been insalubrious, they can hardly have been very pleasant to drink, especially in the summer. It was known, at least as early as Greek times, that pure water is an essential ingredient for a healthy life, but it is also enjoyable to drink. To this day, the Arabs of the desert appreciate clear cold water, and can distinguish between different varieties in much the same way as a wine connoisseur can tell one vintage from another. We should not assume, therefore that the people of Egypt and Mesopotamia were content with their water supplies; they had no alternative.

Before the beginning of the classical period, water-supply systems had been undertaken in areas favourable to their installation, i.e. where supplies of good water were relatively close to urban communities, to which they could be conducted by gravity flow. In prehistoric Europe, where timber was abundant, water was conducted in channels made from planks or through hollowed-out tree trunks. Open terracotta conduits were used on the mainland of Greece by the Mycenaean builders of Tiryns and Philacopi, but the Cretans of Knossos had preferred closed ducts to prevent the accumulation of silt. The palace at Knossos contained very advanced water pipes, each section being about 76 cm long, and from 8.5 to 17 cm in diameter, with a wall thickness of 2 cm. They were slightly tapered towards one, the wider end being flanged.[20]

The oldest, and most extensive waterworks in any ancient city, were those of Jerusalem, where the original works date to about 1000 BC, and are traditionally ascribed to King Solomon. Jerusalem's sources of water were springs south of the city, and the available water was collected in three large artificial reservoirs before being channelled to the city. Within Jerusalem, water storage by means of tanks and cisterns was commonplace. Initially, such reservoirs were adapted for natural formations, but ultimately wholly artificial ones were built, their roofs carried on a forest of pillars. What began in Jerusalem was later applied on a grand scale in Byzantium. The aqueducts of Jerusalem are of imprecise date and uncertain, mainly because Roman reconstructions and renovations have made it difficult to identify the earlier works. One particularly interesting possibility, however, is that there was an inverted siphon in one of the lines; 15-inch stone blocks pierced with a 2-inch diameter hole have been found. A typically Palestinian technique was the *sinnor*, basically a tunnel cut from inside a city to reach an outside well, thus ensuring the water supply when the city was under siege. The *sinnors* at Jerusalem, Gezer and Megiddo are all very ancient, possibly dating before 1000 BC, although in their final form, with carefully graded tunnels to make the water flow by gravity into the city, they are somewhat later.[21]

In 703 BC the Assyrian king Sennacherib began a series of water-supply projects to bring water to his capital Nineveh, near present-day Mosul. In that year he began to draw water from the River Khosr to augment the flow of the Tigris at Nineveh and so ensure the water supply of the city, and make the irrigation of its orchards easier. A diversion dams across the Khosr fed water into a ten-mile-long canal, which was 'dug with iron pickaxes'. By 694 BC supplies were again insufficient, so from the mountains north-east of Nineveh 18 small rivers and streams were canalised into the Khosr and two new dams were built so that the greatly increased flow of the river could be diverted to the city. Four years later even more water was required, so in 690 Sennacherib embarked upon the most ambitious project of all. A masonry dam was constructed on the Atrush river at a place called Bavian, about 30 miles from Nineveh, where the river ran through a deep gorge. From these headworks Sennacherib's engineers dug a winding canal, which was 36 miles long and discharged into the Khosr. Substantial traces of these works can still be seen today. The most impressive is the Jerwan aqueduct bridge, a structure 300 metres long and 12 metres wide which carried the Atrush canal across a valley.[22]

The best known of Greek water-supply systems was that of Pergamon. It was built during the reign of Eumenes II, about 200–190 BC. Beginning far to the north-east of Pergamon in the Madaras-dag, much of the aqueduct line operated by gravity flow. But Pergamon itself stood on very high ground, so that to reach the city the aqueduct in its final 2,000 metres had to cross two deep valleys separated by a ridge before climbing some 200 metres to the final delivery point. Both crossings were accomplished with inverted siphons; an examination of the profile of the route leads us to the conclusion that the pressures at the lowest points must have been of the order of 15 atmospheres. Another well-known example of Greek water-supply engineering is the aqueduct of Samos, described by Herodotos as follows:

The Samians are the makers of the three greatest works to be seen in any Greek land. First of these is the double-mouthed channel pierced for 150 fathoms through the base of a high hill; the whole channel is seven furlongs long, eight feet high and eight feet wide; and throughout the whole of its length there runs another channel 20 cubits deep and three feet wide, where the water coming from an abundant spring is carried by its pipes to the city of Samos. The designer of this work was Eupalinus son of Naustrophus, a Megarian.

The tunnel itself is 1,100 metres long, but though started from both ends hardly deviates from a straight line.[23]

In the next section we shall be describing the *qanat*, a sophisticated method for tapping and delivering underground water. This technique was developed in Assyrian times, so we can see that the whole array of techniques for supplying water — dams, reservoirs, aqueducts, tunnels and qanats — had been introduced before Roman times, so Roman engineers could draw upon this accumulated knowledge when they constructed the water-supply systems for Rome and many provincial cities. In fact, they inherited more than simply a body of knowledge, since they also employed a number of Greek engineers in their hydraulic undertakings. The achievements of the Greeks and other peoples in this field have tended to be overshadowed by those of the Romans; writers have devoted much attention to Roman waterworks and there is now a large number of books on the subject. This interest can be accounted for partly by the sheer scale of the undertaking, and partly by the quality of the surviving evidence. Frontinus was appointed to the post of water commissioner in AD 97 and has left us a valuable record of his administration of the installations, together with a good deal of information about the construction and maintenance of the system.[24] Furthermore, many remains of aqueducts, reservoirs and conduits still exist, and these have been examined and recorded by several scholars. While we should give credit to those who developed the techniques in earlier times, there is no gainsaying the magnitude of the Roman achievement: the only successful large-scale water-supply installation before modern times.

The earliest of the aqueducts supplying Rome was the Appia, constructed in 312 BC, and others were built as the population grew, until by the time of Frontinus there were nine, and another three were built after his death. The total length of these aqueducts was in excess of 300 miles, of which only about 40 miles were carried on arches. Some aqueducts followed very similar routes across the Campagna, and more than one aqueduct sometimes shared the same bridge. In fact, only about 5 per cent of the total length of Rome's aqueducts were carried on bridges. It is worth emphasising this point, since some writers have propounded the idea that the Romans indulged in structural displays for their own sake.[25]

In all parts of the Roman Empire, strenuous efforts were made to provide urban populations with adequate supplies of potable water and it would be impossible in the space available even to list the systems in Spain, France, North Africa and the East. The construction of some of the more remarkable of the aqueduct bridges will be discussed in Chapter 4. Some of these have survived more or less intact, and the beautiful bridge at Segovia is still in use. Writing about AD 1154, the Arab geographer al-Idrīsī mentions several Roman aqueduct bridges that were still in use, including those at Almunecar and Toledo.[26]

Although the basic array of techniques was available to the Romans from the outset, they made some notable innovations to water-supply technology themselves. Their large-scale use of water-resistant cement not only enabled them to build more elegant structures for a lower outlay in labour and materials, but was also used extensively in the construction and lining of water channels. Another innovation is the concept of storing potable water behind dams. Storage in reservoirs inside towns, and the use of diversion dams for irrigation were both old techniques, but impoundment at source for water supply was something new. The three outstanding examples of storage dams are to be found in Spain, two at Merida and one near Toledo. The first two are still in use. A third technique, which had been known earlier, but never applied on a large scale, was the use of the inverted siphon to cross deep valleys, in cases where a bridge would have had to be very high, and hence expensive to construct. The nine siphons in the aqueducts supplying Roman Lyon with water are the most notable examples of this technique for which we have firm evidence. Each siphon might contain up to a dozen parallel pipes, and an enormous quantity of lead was required for a single crossing; it has been estimated that the large Beaunant siphon on the Gier aqueduct at Lyon contained at least 2,000 tons of metal.[27]

The size of the Muslim world in the Middle Ages makes any attempt to generalise about the type and size of water supply systems in Muslim towns and cities impossible. Furthermore, many cities experienced a cycle of birth, growth, maturity and decay, so that to give a reasonably accurate picture of any aspect of life in a given place, it is usually necessary to specify the period under consideration fairly closely — no more than a century, sometimes less. We also know less about water supply in pre-Islamic times than we do about irrigation, although there is evidence than some earlier installations were taken over intact, and continued in use. This applies, for example, to the Roman aqueducts in Spain, mentioned above, and to the waterworks at Ahwaz, in the Iranian province of Khuzistan. A great masonry dam, provided with sluice gates, had been built across the river Karun in Sasānid times. It served to raise the water to the level of the city and ensured that there was abundant water for the needs of Ahwaz and for irrigating its lands. Many water-wheels raised the water into aqueducts, through which it flowed into reservoirs in the city. The dam, called the *shadhurwān*, also provided water for a number of powerful mills. The best description of the system occurs in a work by the great geographer, al-Muqaddasī, written between AD 985 and 990. He says, 'They say that if it were not for the shadhurwan Ahwaz would not be occupied and its rivers would be useless.'[28]

We can allow ourselves one generalisation about Muslim towns and cities in their times of prosperity, namely that they used a great deal of water, not only for drinking and domestic purposes, but also for industrial uses, especially textiles, and for public baths and fountains. In AD 993 the baths in Baghdad were counted and found to number 1500,[29] and many other examples could be given for baths in cities and towns, and fountains in public and private estates. In many cases we are not told how the water was stored and distributed, but simply that the canals ran into the city, then through the streets and into the mosques, dwellings and gardens. There is sufficient information, however, for us to regard the typical system as being the conducting of water by stream, canal or qanat into the city, where it was stored in cisterns. Conduits from these cisterns, often underground, led to the various quarters, and into residences, public buildings and gardens. The surplus water flowed out of the city into the irrigation system.[30] The Muslims were concerned with the maintenance of supplies of good water to sustain life and nourish the crops, but not solely for those purposes. When in Damascus, al-Muqaddasī exclaims, 'there are no baths more beautiful, no fountains more wonderful'.[31] Anyone who has visited Granada, and seen the palace called the Generalife, built early in the fourteenth century, will appreciate the Muslims' regard for water:

With the constantly renewed trees and flowers and the flowing and bubbling of its water, the Generalife evokes, even more than the Alhambra, the private life of the Nasrid princes. And the architects of Granada have never surpassed this perfect alliance of gardens, water, landscape and architecture, which was their supreme aim, and sets the seal upon their art.[32]

In the eastern part of the Roman Empire the city of Constantinople, founded by the emperor Constantine in AD 330, where stood the old Greek city of Byzantium, was to be the capital of an empire that survived until AD 1453. It remained a great power even after its fairest provinces — Syria and Egypt — were lost to the Arabs in the seventh century. The lands of the Byzantine Empire, as it is known to history, were further attenuated by the occupation of most of Asia Minor by the Seljuk Turks in the eleventh century, and by the seizure of most of its territory, including Constantinople, by the Latins in the thirteenth, in the disgraceful episode called the Fourth Crusade. Nevertheless, the Byzantine Empire remained a force to be reckoned with until the capture of Constantinople by the Ottoman Turks in 1453. However, except in the fields of theology, church architecture, politics, administration and art we have very little information remaining from over 1,000 years of Byzantine civilisation. The earliest elements of Constantinople's water-supply systems are believed to date to the reign of Hadrian, with additions in the time of Valens in the fourth century, and more substantial constructions in the reign of Justinian in the sixth century. Most of the three dozen or so bridges that carry Constantinople's aqueducts are attributed to Justinian. Over a period of many centuries the city frequently had to withstand sieges, and it is therefore not surprising that measures were taken to ensure the internal water supply. Defensible bodies of water were constructed in the form of a number of interlinked storage tanks, some of them quite small, but several of them very large, so that they resembled the naves of drowned cathedrals, because of the forest of columns that supported their roofs.[33]

Roman standards were maintained in Islam and Byzantium, but in northern Europe there was a drastic decline. In some areas, notably in Italy, the Roman installations continued to be used, but in general water supply reverted to the pre-Roman situation. Or perhaps we should add the qualification that large parts of the region that had been nominally part of the Roman Empire had never been thoroughly Romanised. This was certainly the case in northern Britain, parts of France and western Germany, and even in the more remote areas of Italy itself. In the Middle Ages water supply in Europe was from wells, streams and rivers, and in the small but crowded towns of medieval times impure water and bad sanitation were major factors in the spread of epidemic diseases. In the twelfth and thirteenth centuries efforts were made to improve the water supply of Paris, and throughout medieval times monastic establishments took steps to provide pure water for their own purposes. Only after the Renaissance, however, were serious attempts made to improve the water supply of European cities.

Achievements of Early Hydraulic Engineering

For the remainder of this chapter we shall be considering some aspects of hydraulic engineering that were of crucial importance to society and to the development of modern hydraulic systems.

Qanats

The *qanat* is an almost horizontal underground conduit that conducts water from an aquifer to the place where it is needed. The qanat should not be confused with tunnels or covered aqueducts, which carry water from above-ground sources such as streams or lakes; the technique for constructing qanats is quite different and highly specialised. In Iran, where qanats are still an important source of water, the construction of qanats is in the hands of experts (*muqanni*), and the secrets of the profession are largely handed down by word of mouth from father to son. The construction of a new qanat requires a considerable outlay of capital and there is always the risk that financial returns may be low if the eventual flow of water is inadequate. It is customary, therefore, for the landowner or other authority to engage a skilled surveyor, usually a former muqanni with great field experience and keen powers of observation, for the preparatory work. The termination of the qanat, either farmland to be irrigated or a community to be provided with potable water, or both, will be known in advance, as will the general location of likely aquifers. The surveyor then carefully examines the alluvial fans, looking for traces of seepage on the surface and often for a hardly noticeable change in vegetation, before deciding where the trial well (*gamana*) is to be dug. A winch is set up at the selected point, and two muqanni begin to excavate a vertical shaft about three feet in diameter, the spoil being hauled to the surface by two labourers and deposited around the mouth of the shaft. When the muqanni reach the aquifer they proceed slowly until they reach the impermeable layer. The well is then left for a few days, during which time the water is hoisted up in leather buckets, and its quantities noted, while at the same time any fall in the level of water is observed. The surveyor then has to decide whether they have reached genuine groundwater, or just some water trickling in from a local clay or rock shelf. If necessary more trial wells are dug to find a genuine aquifer or to determine the extent of the one already found and its yield. The shaft with the highest yield is chosen as the 'mother well'; in some cases all trial wells are linked together with a conduit, thus forming a water-yielding gallery.

The next step is for the surveyor to determine the route, gradient and precise outlet of the qanat. The route will be selected according to considerations of terrain and, in some cases, questions of ownership. To start the survey, a long rope is let down into the mother well until it touches the surface of the water, and a mark is made on it at ground level. The surveyor then selects a spot on the route 30 to 50 yards from the mother well for the first ventilation shaft. A staff is held on this spot by a labourer, and the surveyor measures the fall with a level. Nowadays a modern surveying level may be used but in earlier times one of the simple instruments described in Chapter 7 was used. A second mark is made on the rope coinciding with the measurement on the staff; the distance from this mark to the lower end of the rope will be the depth of the first ventilation shaft. He continues to level in this way along the route, marking the rope at the location of each shaft, until he reaches the end of the rope. He has then reached a point on the ground level with the surface of the water in the mother well. For the mouth of the qanat he now chooses a place below this level, but higher than the fields, and divides the drop from the level point to the mouth by the number of proposed ventilation shafts and adds this amount to the previously surveyed depth of each shaft. In this way he determines the gradient of the conduit, which is usually from 1:1000 to 1:500.

After completion of the survey, a number of guide shafts, about 300 yards apart, are driven under the supervision of the surveyor. Then the rope with the marked length of each vertical shaft is handed over to the muqanni. He now begins to work with his assistants by driving the conduit into the alluvial fan, starting at the mouth. At first the conduit is an open channel, but it soon becomes a tunnel. Another team sinks ventilation shafts ahead of the tunnellers, and labourers haul up the spoil to the surface through these shafts. The tunnel is about 3 feet wide by 5 high, and work can proceed fast through reasonably firm soil, but in soft, friable soil the roof is unsafe, and hoops of baked clay, oval in shape, have to be used as reinforcement. Two oil lamps are kept burning on the floor of the conduit; sighting along these, the muqanni keeps the tunnel in alignment, and they also serve as a warning of poor air, since they go out before there is danger of a man suffocating. As the work nears the mother well great care has to be taken in case a muqanni misjudges the distance and strikes the full well, in which case he might be swept away by the sudden flow (Figure 2.2).[34]

The flow of qanats varies so greatly that it is difficult to give an average figure, but 10 to 20 gallons a second can be regarded as normal. Recent estimates have shown that 75 per cent of all water used in Iran comes from qanats and that their total length exceeds 100,000 miles. The city of Teheran alone has 36 qanats, all originating in the foothills of the Elburz 8 to 16 miles away, with a measured flow of 6.6 million gallons a day in spring and never below 3.3 million in the autumn. The qanat may therefore be considered as one of the most successful of man's inventions, since it has been in continuous use for over 2,500 years. Outside Iran qanats are still in use in parts of the Arab world, notably in the south-east of the Arabian peninsula and in North Africa.[35]

The Assyrian king Sargon II (reigned 722–705 BC) learned the secret of tapping underground water during his conquest of Urartu

Figure 2.2: Qanat

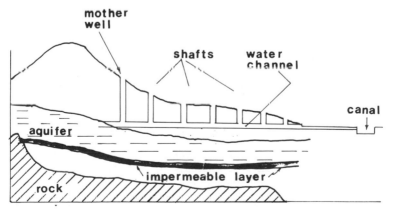

(present-day Armenia). It is probably no coincidence that Armenia was one of the oldest mining and metallurgical centres in the Middle East, since the technique of qanat-building is very similar to that of mining. The method of disposing of surplus water through horizontal adits may well have inspired the idea of the qanat. In the ancient world qanats were known in Achaemenid Iran, in Arabia and in Egypt.[36] They were in widespread use in the Muslim world, from the Middle East through North Africa to the Iberian peninsula. In post-Columbian times they were introduced to Central and South America. Arabic writers often refer to qanats in connection with both water supply and irrigation. The great city of Nishapur in Khurasan was supplied entirely by qanats; most of these ran under the city and surfaced outside to form a river which irrigated the gardens of the area and many of the rural cantons. Some qanats, however, surfaced inside the city and provided water to buildings and gardens. The irrigation system was supervised by a force of inspectors and guards, who were also responsible for the maintenance of the qanats.[37] The city of Rayy, near the site of modern Teheran, also obtained its water for drinking and irrigation from qanats.[38] At the other end of the Muslim world the city of Tangier was supplied by qanats from a great distance.[39]

It is worthwhile recording the existence of well-preserved remains of a qanat system near the ruins of Palmyra in Syria, which was examined briefly by the present writer in 1952. In the low hills about two miles to the north of the ruins it is possible to descend to a tunnel that is only a few feet below ground level. The tunnel is constructed of unmortared masonry, of such good workmanship that it is impossible to insert the blade of a small knife between the blocks. The cross-section of the tunnel is about 3 feet wide by 4 feet high, with an arched roof. An interesting feature is the zigzagging of the tunnel, pre-sumably to cut down the speed of flow. It was not possible to trace the qanat — if such it is — back to its source, which may have been an aquifer somewhere in the hills. This is probably not an Islamic construction, but may date back to Roman times, or to the period of Palmyra's independence in the first three centuries of our era.[40]

Roman Water-supply

Estimates of Rome's size and population at different times, reached by various methods, suggest the following numbers: about 400,000 at the beginning of the first century BC; over half a million by about 50 BC; approaching one million in the first century AD and reaching a peak at something over one million in the middle of the second century AD, and thereafter declining.[41] The first aqueduct, the Annia, was built in 312 BC and others were added to keep pace with the increasing population, with the major period of aqueduct construction from about 100 BC to AD 150. The conduits consisted of channels, all of them at least two feet wide and, with one exception, at least five feet high. Each was large enough to allow workmen to enter and carry out repairs, and the water did not, of course, fill the conduit completely. The channels were lined with water-resistant concrete and covered to protect the water from sunlight. This type of conduit, just below ground level, was certainly the cheapest method of construction and the Romans, being a practical and economically-minded people, did not resort to other methods unless they had to. If there will hills between the fountainhead and the city they were pierced by tunnels, unlined if the formation was of tufa or stone, but otherwise the floors and walls were made of masonry and the roofs were vaulted. Vitruvius recommends that there should be vertical shafts at intervals of 120 feet and that the usual fall of 1:200 be maintained through the tunnels.[42]

The inverse problem, that of crossing valleys, was solved either by erecting aqueduct bridges, or by the use of inverted siphons. An inverted siphon is simply a pipe that runs from a head tank on one side of the valley to a crossing (usually a bridge) at the valley bottom and then up to a receiving tank on the other side. The difference in level between start and finish must be large enough to give the water velocity and to overcome the frictional resistance in the pipe (see Figure 2.3). Norman Smith has published a critical study of the Romans' use of inverted siphons, and of a number of misconceptions about the technique in the works of nineteenth- and twentieth-century writers.[43] Dr Smith points out that the Romans went to great lengths (literally) to avoid valleys, but that when a crossing was necessary siphons were preferred to bridges if a bridge would have been more than 150 200 feet high. Presumably siphons were cheaper than bridges for the deeper crossings.

There is a good deal of archaeological evidence for the use of siphons, but the only written source is Vitruvius. The description of siphons by Vitruvius is obscure, and this has caused difficulties for modern writers, who have usually regarded Vitruvius as a reliable source. This may not be the case: we have no other work with which to compare *De Architectura*, but it is certainly an uneven work. Much of the information it contains is valuable, but some passages are obscure

Figure 2.3: Inverted Siphon

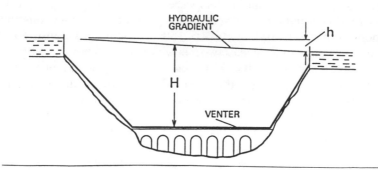

and others show clearly that Vitruvius did not really understand the techniques that he was describing. One of the remarks made by Vitruvius has led modern writers into difficulties, since he implies that valves are needed in siphons to relieve the air pressure. But there is no air pressure in such pipes, only the hydraulic pressure caused by the differential head of water. Doubts have also been raised as to whether lead pipes could stand this pressure, but Smith quotes the specification of one of the siphons of the Gier aqueduct, built during the reign of Hadrian to supply Lyon, the last of four aqueducts built for that city. The Soucieu siphon had the following specification:

External pipe diameter	10 inches (0.26 m)
Number of pipes	9
Length of siphon	3,950 feet (1204 m)
Fall	29 feet (8.844 m)
Maximum depth	304 feet (92.822 m)

The tensile strength of lead is about 2,000 pounds per square inch, and a 10-inch pipe with a wall thickness of $^3/_4$ inch would be stressed to about 1,000 pounds per square inch under an operating head of water of 400 feet, adequate for any of the Lyon siphons. Pipes of this type have much greater wall thicknesses than those described by Vitruvius, but he was describing pipes for normal distribution and transmission lines. A reliable estimate for the total weight of lead in the nine siphons in the aqueducts serving Lyon is from 10,000 to 15,000 tons. Siphons were therefore only used when there was no more economical alternative, since the mining, processing, transporting, fabricating and laying such large quantities of lead must have taxed even the Romans' resources. On the general design of siphons, Dr Smith has the following comments to make:

> The essentially straightforward position was have arrived at then is this. A Roman siphon comprised a run of parallel lead tubes connecting a header tank and a receiving tank each at the end of the conventional open channel. Valves in the venter were used to release air when the flow was started up, or they could be opened to drain the siphon when necessary; otherwise they were permanently closed. This specification will not explain all of Vitruvius but nowhere is it seriously at odds with him, and it does conform to the extant archaeological evidence.[44]

A typical flow diagram from the source to the terminations in the city is shown in Figure 2.4. The water was first led into settling tanks, which were lined with concrete and had sloping floors to facilitate cleaning. These tanks were often in pairs, so that one could be cleaned while the other remained in service; there might alo be intermediate settling tanks at intervals along the route. The aqueduct proper terminated in a collecting tank, whence it was conducted to the city by

Figure 2.4: Roman Water-supply System

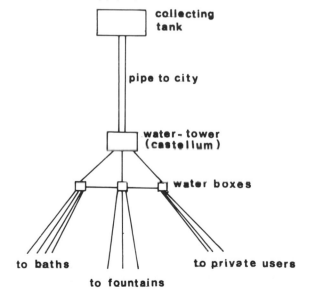

pipes that discharged into a water tower (*castellum*). From the castellum the water was piped to three water boxes; the outer two were connected to the central one by pipes so that they overflowed into it when the water reached a certain level. From the central water-box water was distributed for public uses, including cisterns and fountains, military barracks and public buildings. This supply was the most important purpose of the Roman system, and had priority over all other uses. One of the two outer boxes supplied the baths and the other supplied private users — both domestic and industrial. The public could draw water from the cisterns and fountains without payment, but a tax was levied on supplies tapped from the mains to private premises. The Emperor's permission was required for any private installation. If this was granted a bronze nozzle stamped with an official mark was fitted to the main and a pipe from the nozzle took the water to the owner's premises. There were a number of standard nozzle sizes, based upon the Roman digit. There were about 15 nozzles in common use, the smallest being 2.31 and the largest 29.57 cm in diameter. Lead pipes were also made in standard sizes: Vitruvius lists ten sizes, not expressed in terms of the internal diameter as is the modern practice, but derived from the width of the sheet of lead used, that is, the circumference plus the overlap for soldering. From these figures we can calculate that the smallest pipe had an internal diameter of about 1.32 cm and the largest an internal diameter of 57.4 cm.

According to Vitruvius' specification, all these pipes would have had the same wall thickness of about 0.63 cm and since the tensile stress in the pipe walls is directly proportional to diameter, the largest pipe would have been weaker than the smallest in the ratio 57.4:1.32. Furthermore, it would also have been very liable to damage from external loads.[45] Frontinus tells us that various types of fraud were practised, either to avoid paying water-tax or to obtain more water than was stipulated. For example, secret tappings were made into the public mains or the nozzles were larger than the specified diameter. Nevertheless, the Roman system of pipes and nozzles is the first known example of large-scale standardisation of components, which is an indispensable requirement of modern engineering.

The total length of the aqueducts serving Rome was over 300 miles, and there were also extensive systems in the provinces. The maintenance of the channels and their substructures was always a serious problem. Frontinus is often unsound on technical matters, but his brief discussion on the upkeep of the aqueducts is quite reliable.[46]

He says that damage could be caused by the lawlessness of abutting proprietors, by age, by violent storms and by defects in the original construction. He is not implying, in his use of the word 'lawlessness' that wilful damage was done, but that the statutory clearway for 15 feet on either side of the channel had been encroached upon. The roots of trees, in particular, could cause damage to underground conduits. Another cause of damage was the accumulation of mineral incrustations. Frontinus was also aware that the exposed sections of the aqueducts were more liable to suffer from the effects of heat and frost than were the underground sections. He recommends that repairs to the channels themselves should not be carried out in the summer, when people had most need of water, and that only one aqueduct should be taken out of commission at any one time. It was possible, however, to repair a section of a channel without stopping the flow, by building a temporary bypass with lead channels. Masonry work was to be repaired in the period from April 1 to November 1, with the proviso that it should be interrupted in the hottest part of the summer, because 'moderate weather is necessary for the masonry properly to absorb the mortar and to solidify into a solid mass; for excessive heat of the sun is no less destructive than frost to masonry'. This is perfectly sound advice.

The writings of both Frontinus and Vitruvius are a rather curious mixture of reliable technical knowledge, such as the passage quoted above, and passages full of obscurities and even complete nonsense. Not suprisingly perhaps, they appear at their best when discussing straightforward constructional matters, and at their worst when the subject demanded at least an empirical understanding of natural phenomena. There is a well-known passage in Frontinus, for example, which indicates that he believed that the rate of flow through a conduit depended only upon its cross-sectional area, not also upon the speed of the current.[47] Some errors may, of course, have entered the texts of these two writers as they were copied over the centuries, but the possibility remains that at times they did not know what they were talking about. It should not be assumed, however, that the engineers who actually planned, surveyed and supervised the construction of the aqueducts were ignorant of the principles of hydraulics. We must assume that they had access to an accumulated store of empirical data. It is difficult to imagine, for example, that they could have equalised the flow, in a long aqueduct, between the sections consisting of open channels and the pressurised siphons without a set of empirical rules to guide them.

Hydraulic Works in Islam

We can obtain a clear picture of the importance of irrigation works in Islam by consulting a number of sources, notably the works of Muslim geographers. The *Kitāb al-hāwī*, an anonymous Iraqi treatise of the early eleventh century, also gives us valuable information about the construction and maintenance of irrigation works.[48] In a section devoted to quantity surveying, which shows that methods have not changed materially over the centuries, instructions are given for calculating the quantities of earth to be excavated from canals of given lengths, widths and depths and for converting these quantities into manpower requirements. The canal banks were reinforced with bundles of reeds, and the man-hours required for preparing and placing the bundles are given. For excavation, the number of diggers (called 'spades') was first calculated, and to these were added the number of carriers to each spade, which depended on the distance the spoil had to be transported. Overheads for ancillary workers and supervision were then added. There was a set price for each task, so in the end a Bill of Quantities was produced which would provide the estimate for the cost of the works and serve as a guide for the recruitment of labour. Or, if the project was let out to subcontract, which was often the case, the Bill of Quantities would be the main document for awarding the contract and for the subsequent measurements and payments.

From this document, and elsewhere, we get a picture of a highly organised State enterprise, with an army of bureaucrats, engineers and surveyors, controlling a very large labour force, whose productivity and rates of pay were closely specified. Although the calculations in the *Kitāb al-hāwī* are only 'worked examples', they were probably based upon typical jobs done by the department in the past. The numbers of men resulting from the calculations are between 500 and 1,600, and there must have been a number of projects of this size in hand at any given time. We also have reports such as the one quoted above for the district of Marv, which gives the labour force for the irrigation department as 10,000 men.[49]

The superintendent of irrigation, in addition to supervising construction and maintenance, had the reponsibility for ensuring that the available water was allocated equitably. One method for apportioning water, which was used, for example in the Ghuta of Damascus and in parts of Spain, was by proportion. A river was considered to be divided into successive stages, each stage representing the point of derivation of one main canal which drew all the water at that stage, or of two canals, dividing the water between them. At each stage the river was considered to hold 24 units of water. To quote Thomas Glick:

> The units were not, however, expressed in hours, but as simple proportions of a whole. Thus, in times of abundance, each canal drew water from the river according to the capacity of the canal; in times of drought, the canals would take water in turn, for a commensurate number of hours or a proportional equivalent. The same was true of individual irrigators (and herein lies the genius of the Valencia system): when the canal ran full, each irrigator could open his gate as he pleased, but when water was scarce, a turn was instituted; each irrigator, in turn, drew enough water to serve his needs (this style of irrigation was by submersion of the field, typically to a standard depth of an ankle). But he could not draw water again until every other irrigator in the system had his turn. Thus a relatively equal distribution was ensured, both in times of abundance and of scarcity, and no measurements of time or orifice of delivery were needed.[50]

In other areas, for example the Yemen and parts of southern Spain, allocation of irrigation water was by fixed time units which could be bought and sold. One method of measuring the time units was the *tarjahār*; this was a metal bowl, having a small orifice in its underside. At the moment when the water was let into an irrigator's field the bowl was placed on the surface of the water in a nearby pool. When the bowl sank the water was cut off by closing the small dyke at the mouth of the irrigator's supply channel.[51] Another method of apportioning water was analogous to the Roman system of providing nozzles of approved sizes for domestic consumers. The irrigators' supply channels were closed at the entrances by boards in which there were a number of holes of stipulated diameter.[52] The famous Nilometer on the island of Roda on the Nile represents a very special example of the measurement of irrigation water, mainly for the purpose of taxation. This consists of a tall, graduated octagonal column, standing in a stone lined pit, which is connected to the river by three tunnels at different levels. The pit is provided with a staircase. The Nilometer, in its present form was constructed in AD 862 or 863, and although there have been repairs and alterations since then, it remains essentially as it was when Ibn Jubayr examined it in the twelfth century. According to his account, which agrees in most respects with that of al-Idrisi, the column was divided into 22 cubits, each subdivided into 24 fingers. A rise to the nineteenth cubit was good, 17 was average, and no tax was levied if the water did not reach the sixteenth cubit.[53]

The point has already been made that it is usually not possible to separate the water supply from the irrigation systems when considering hydraulic works in Islam, since the main feeders — canals or qanats — were used for both purposes. There is no parallel to the Roman system of aqueducts, designed almost solely for water supply, for any Islamic city. Aqueduct bridges were fairly common, however; in Spain the Roman structures often remained in service, but there are cases where new bridges were built. Water was conducted to the city of Samarkand, for example, in a lead-lined channel carried on a bridge. This was necessary because the land around the city had been excavated to provide clay for the building of the city.[54]

Inside the cities, the arrangement for the storage and distribution of water seem to have been much the same as they were for Rome, although we have no source similar to Frontinus to give us a detailed description. Requirements were similar: public baths, drinking fountains, supplies to public and private buildings and industrial uses. In addition, there was the requirement, specific to Islam, for the provision of houses for ritual ablutions. In the twelfth century there were 40 of these in Damascus, which was not then a very large city.[55] Baths were often of the 'Turkish' variety. Writing early in the fourteenth century, the Egyptian writer Ibn al-Ukhuwwa describes the bath as having three chambers: 'The first chamber is to cool and moisten, the second heats and relaxes, the third heats and dries.'[56] The baths and the supply tank had to be thoroughly cleaned daily. Hot baths (*hammām*) were in use in the Muslim world from the seventh century onwards; in the tenth century water was piped from the hot springs in Tiberias to the city's baths.[57] At the other extreme of temperature, in the same period, there were said to be over 2,000 points for dispensing iced water in the city of Samarkand.

In general, the Muslims were as concerned with obtaining supplies of good water as were the Romans. Indeed in some instances, Roman works, having fallen into decay, were restored to their former condition after the Arab conquests. A good example of such a restoration is the city of Qayrawan in Tunisia. This was founded by the Arabs in AD 670 at a location once occupied by a Roman or Byzantine town. The older hydraulic system was rebuilt, and later two impressive linked cisterns were constructed for receiving the waters of the Wadi Merj al-I il when it was in flood. These cisterns, which still stand, were completed in AD 862. Although they appear to be circular, both are actually polygonal, the larger having a diameter of just under 130 metres, the smaller one a diameter of 37.4 metres. The smaller receives the waters of the wadi and acts as a settling tank; a circular duct, several metres above its base connects it to the larger cistern, which has a depth of about eight metres. On leaving the larger cistern, the water is decanted a second time into two oblong covered cisterns.[58]

Notes

1. Norman A.F. Smith, *Man and Water* (Peter Davies, London, 1975), p. 7.
2. Raf van Laere, 'Techniques Hydrauliques en Mésopotamie Ancienne', *Orientalia Lovaniensia Periodica*, University of Leuven Press, 11 (1980), pp. 11-53, 24-7.
3. Qur'ān, Sūra 34, 15-21.
4. W. Montgomery Watt, *Companion to the Qur'ān* (George Allen & Unwin, London, 1967), pp. 196-7.
5. Smith, *Man and Water*, p. 16.
6. Ibid., p. 7.
7. Thomas F. Glick, *Irrigation and Society in Medieval Valencia* (Harvard University Press, 1970), p. 176.
8. Thomas F. Glick, *Islamic and Christian Spain in the Early Middles Ages* (Princeton University Press, 1979), p. 77.
9. Ibn Hawqal, *Kitab Surat al-'Ard*, Arabic text ed. J.H. Kramers, 2nd edn of vol. 2 of BGA (Brill, Leiden, 1938), p. 174.
10. Glick, *Islamic and Christian Spain*, p. 73.
11. Robert Ignatius Burns, *Medieval Colonialism*, (Princeton University Press, 1975), p. 122.
12. Guy le Strange, *The Lands of the Eastern Caliphate* (Frank Cass, London; 1st edn 1905, this edn 1966), pp. 28-9.
13. Ibn Sarabiyūn (also called Ibn Serapion) *Kitāb Ajā'ib al-aqālīm al sab'a*, ed. H. von Mžik (Vienna, 1929), pp. 114-38.
14. Raf van Laere, 'Techniques Hydrauliques', p. 22.
15. Al-Balādhurī, *Kitāb futūh al-buldān*, ed. M.J. de Goeje (Brill, Leiden, 1866), pp. 345-71.
16. Al Istakhri, *Kitāb al-masālik wa'l-mamālik*, ed. M.G. al-Hīnī (Cairo, 1961), p. 57.
17. Ibn Hawqal, *Kitāb Sūrat*, pp. 635-6.
18. Al-Muqaddasī, *Ahsān ul-tuqāsīm fi ma'rifat al-aqālīm*, ed. M.J. de Goeje in vol. 2 of BGA (Brill, Leiden, 1906), p. 279, pp. 280ff. See also Le Strange, *The Lands of the Eastern Caliphate*, pp. 460-73.
19. Smith, *Man and Water*, pp. 69-70.
20. R.J. Forbes, *Studies in Ancient Technology* (6 vols., 1st edn Brill, Leiden, 1957-8), vol. 2, 2nd edn 1964, p. 153.
21. Smith, *Man and Water*, p. 77.
22. Forbes, *Studies*, vol. 1, pp. 160-3.
23. Ibid., p. 164.

24. Frontinus, *The Strategems and the Aqueducts of Rome* (1st edn Loeb Classics, London, 1925; this reprint 1980), the *Aqeducts*, ed. Mary B. McElwain.

25. Smith, *Man and Water*, p. 82.

26. Al-Idrīsī, *Description de l'Afrique et de l'Espagne*, Arabic text with French trans., eds. R. Dozy and M.J. de Goeje (Brill, Leiden, 1866), pp. 187, 199 in Arabic, pp. 228, 242, Fr.

27. Smith, *Man and Water*, p. 91.

28. Al-Muqaddasī, *Ahsān al-tāqasīm*, pp. 411-12.

29. A.A. Durī, 'Baghdād' in EI, vol. 1, 1960, p. 899.

30. Such, for example, was the arrangement in the city of Zaranj, capital of Sijistān (now western Afghanistan), according to Ibn Hawqal, *Kitāb Sūrat*, p. 414; similarly in Nisibīn in northern Syria — see Ibn Jubayr, *Rihla*, Arabic text ed. W. Wright (Brill, Leiden, 1852), 2nd ed amended by M.J. de Goeje (Brill, Leiden 1907), p. 239.

31. Al-Muqaddasī, *Ahsān al-tāqasīm*, p. 157.

32. H. Terrasse, 'Gharnāta', in EI, vol. 2, 1965, p. 1019.

33. Smith, *Man and Water*, pp. 94-5.

34. Hans E. Wulff, *The Traditional Crafts of Persia* (MIT Press, Cambridge, Mass., 1966; 2nd printing 1976), pp. 251-4; and Henri Goblot, *Les Qanats; une Technique d'Acquisition de l'Eau* (Mouton Éditeur, Paris, 1979), pp. 28-35.

35. A.K.S. Lambton, 'Kanats' in EI, vol. 4, 1978, pp. 528-32.

36. Lambton in EI.

37. Al-Istakhrī, *Kitāb al-masālik*, p. 145; Ibn Hawqal, *Kitāb Sūrat*, p. 433.

38. Ibid., p. 122.

39. Ibn Hawqal, *Kītāb Sūrat*, p. 79

40. During my visit in 1952 I had no opportunity of making a detailed study. It seemed a good idea, however, to mention this conduit, since I have not seen it referred to elsewhere.

41. Smith, *Man and Water*, p. 84.

42. Vitruvius, *De Architectura* ed. F. Granger (Loeb Classics, London 1934, reprint. 1970, 2 vols.), vol. 2, bk. VIII, Ch. 6.

43. Norman A.F. Smith, 'Attitudes to Roman Engineering and the Question of the Inverted Siphon', in *History of Technology*, eds. A. Rupert Hall and Norman Smith, Mansell, London, vol. 1 (1976), pp. 45-71.

44. Ibid., p. 59.

45. Vitruvius, *De Architectura*, vol. 2, bk VIII, Ch. 6.

46. Frontinus, *Aqueducts*, bk II, chs. 65-89.

47. Ibid., chs. 119-30.

48. Claude Cahen, 'Le Service de l'irrigation en Iraq au début du XIᵉ siècle' in *Bulletin d'études orientales*, vol. 13 (1949-51), pp. 117-43.

49. See note 17.

50. Glick, *Islamic and Christian Spain*, p. 71.

51. Al-Muqaddasī, *Ahsān al-tāqasīm*, p. 357.

52. Ibn Hawqal, *Kitāb Sūrat*, pp. 435-6.

53. Ibn Jubayr, *Rihla*, pp. 54-5; Al-Idrīsī, p. 144 Arabic/p. 173 Fr. see also K.A.C. Creswell, *A Short Account of Early Muslim Architecture* (Penguin Books, London, 1958), pp. 292-6.

54. Al-Istakhrī, *Kitāb al-masālik*, p. 177.

55. Ibn Jubayr, *Rihla*, p. 288.

56. Ibn al-Ukhuwwa, *Ma'alim al-qurba fī ahkām al-hisba*, ed. R. Levy, Arabic text with abridged English trans. (Gibb Memorial Series, London, New Series, 1938), pp. 149 ff.

57. Al-Istakhrī, *Kitāb al-masālik*, pp. 44-5.

58. Cresswell, *Early Muslim Architecture*, pp. 291-2.

3   Dams

Design and Construction

Hydraulic systems may be intended for water supply, irrigation or hydro-power, or any combination of these three. This is as true of early times as it is today, but the three types of system were not introduced simultaneously. Engineering methods were applied to irrigation systems at the dawn of civilisation, to water supply by the seventh century BC, and to hydro-power by about the first century BC, although for the last case we have no firm dating. Dams are required in most hydraulic systems, whatever their purpose, but the functions of dams vary. In wadi irrigation, as we have seen, they are used to trap the floodwaters that result from heavy but infrequent downpours, so that the water-level is raised above that of the surrounding fields, to which it can then be conducted under gravity. For perennial irrigation dams are used to divert water from streams or rivers into the canal network. The impounding of rivers behind dams gives more control over the supply throughout the year. As with wadi irrigation, it also allows the water in the reservoir to be gravity-fed into irrigation and town supply systems. A further advantage, if the water is to be used for hydro-power, is that there is a high, fairly constant head of water, which would not be the case if the river were unregulated.

There are two basic types of dam — gravity and arch. The first, as its name implies, relies upon the weight of the dam to withstand the pressure of the water. Gravity dams can fail at any section by slippage or overturning, if the thrust of the water overcomes the shear and frictional resistance at that section. Failure can also occur if the scouring action of the water undermines the foundations. With gravity dams, it is important to keep the line of thrust (the resultant of the pressure of the water and the weight of the dam) inside the 'middle third' of the cross-section, that is, the middle third with the line connecting the centres of gravity of the sections

47

as its centre-line. If the line of thrust goes outside this limit, tensile stresses will be produced which may cause fracture, because masonry, especially its joints, cannot withstand tension. For additional strength, buttresses are sometimes added to the downstream wall of gravity dams. Arch dams are designed to resist the force of water and silt by horizontal arch action and are adaptable only to those sites where the length is small in comparison with the height and the sides of the valley are composed of good rock to resist the arch thrust at the haunches. With rare exceptions (see below) true arch dams were not built before modern times. Dams in the shape of arches did occur when, for example, the engineers discovered that the best rock lay along a curve. Dams of this shape were not, however, real arch dams; they were of massive construction and were, in effect, gravity dams.

The mathematical analysis of the forces acting on dams was, of course, unknown to the early engineers; their designs were based upon accumulated experience, intuition and sound constructional practice. The selection of a suitable site was clearly of the greatest importance. The general area in which a dam was to be located was determined by hydrological and social factors, but for a dam of any size the precise location depended upon finding a site with good load-bearing strata, preferably rock. More than anything else, the stability of a dam depends upon excavating down to a good stratum, and then ensuring that there is a strong bond between this stratum and the structure. This is particularly true of early dams, since the structures were overdesigned by modern standard, and the weakest point was therefore the foundations. Once the site had been chosen, the height of the dam would have been determined. To do this, a fairly accurate survey of the surrounding area would be needed. The engineers had to take into account two main variables — the height of the dam and the extent of the proposed hydraulic system. The systems — whether for water supply or for irrigation or both — were mostly fed by gravity, so the height of the dam was a major factor in determining the size of the system. But the height of the dam was influenced by other factors: the local hydraulic conditions; topography; safety; and the resources available for construction. As always is the case in the design of engineering projects, the best compromise had to be sought from a number of possibilities. We often find that water-raising machines were erected close to dams. It seems unlikely that these were erected as an afterthought, but rather that their use was included in the planning of the whole hydraulic system.

The selection of the materials of construction was influenced partly by the design of the dam and partly by availability. Earth dams were common and are still in widespread use today. They are perfectly satisfactory for certain kinds of service, provided they have a core of clay or other impermeable material and plenty of overflow capacity, but they are not really suitable for high dams. In certain areas, notably Iraq, earth dams were almost universal; they were (and are) quite adequate for diverting rivers into canal systems and, in any case, the cost of transporting large quantities of stone would have been prohibitive. In other areas, where higher dams were needed, some form of masonry construction was necessary. This could be of dressed stone, mortared or not, random rubble or concrete. Quite frequently, dams were constructed by building two masonry walls with a gap between them, and then filling the gap with cheaper material such as earth or rubble. If the dam was designed to discharge overflow water from its crest, this had to be of stone or concrete, since earth would quickly have been worn away by the action of the water.

Another method of disposing of the water was to build spillways controlled by sluices from the reservoir to the downstream side; sluice-gates in the dam structure itself were also used. But since the main purpose of many dams was to impound water, temporarily or permanently, to use for irrigation or water supply, the water had to be conducted to where it was needed. In some cases this was done by simply digging supply canals from the banks of the reservoir to the start of the hydraulic system. Yet another method was to build a single large conduit from a spillway, through a settling tank, into a distribution tank, from which the supply canals were led out. Underground conduits from the reservoir to a collecting tank were also constructed. The mouths of these tunnels were some distance above the bottom of the reservoir, otherwise they would have been clogged with silt. Silt is a common problem with dams and their associated hydraulic works, but whereas canals can be cleaned regularly, albeit at considerable expense, it was not usually possible to clear the reservoirs of silt. Reservoirs therefore gradually silted up over a period of years, and the amount of water stored progressively decreased.

In the space available, it will only be possible to list the more important dams built in five distinct cultures, with particular reference to dams which embody features of especial engineering interest. In order to provide an historical perspective, we begin with the dams built in pre-classical times.

Antiquity

The earliest known dam was built in Egypt some time between 2950 and 2750 BC, near Helwan, about 20 miles south of Cairo. It is known as the Sadd al-Kafara, which means 'Dam of the Pagans' in Arabic, and its remains have survived to this day. It was built across the Wadi al-Garawi, with a crest length of 348 feet and a base length of 265 feet, with a maximum height of 37 feet above the bed of the valley, and it was immensely thick. It was constructed of two walls of random masonry, each 78 feet thick at the base, containing together at least 30,000 cubic yards of masonry, all of which had to be quarried, transported to the site and put into place. Between the two walls was a gap of 118 feet, and this was filled with some 60,000 tons of gravel from the wadi bed. The reservoir formed by the dam was apparently intended to provide drinking for workers at nearby quarries. It does not seem to have been in use for very long. The dam is very loosely built and would not have been watertight, but nevertheless it is a remarkable example of a major civil engineering project, especially when we consider that it was built over 4,500 years ago. The Sadd al-Kafara had no successors in Egypt until modern times. This is probably because basin irrigation, the only method used in Egypt until recently, needs nothing more than low dams and dykes built of earth.[1]

There is not much firm information about the building of dams in Mesopotamia before 1000 BC, but it is known that they were an essential part of the irrigation systems, and that these included at least one major dam across the Tigris, constructed about 1750 BC. Dams were constructed either of masonry or of reeds and earth combined.[2] The hydraulic works of the Assyrian king Sennacherib, to supply water to Nineveh, have already been mentioned in the previous chapter. Altogether four dams were built, between the years 703 and 690 BC, one over the River Khosr and the others over smaller streams in order to divert their flow into the Khosr and hence increase its discharge. Remains of the second two of these show that they were built of roughly shaped blocks of limestone, sandstone and conglomerate, all firmly mortared together. They were not very high — the third had a maximum height of $9\frac{1}{2}$ feet

and the second was even lower; they were 340 and 750 feet long
respectively. Neither of these dams, nor the fourth, were built
straight across the valleys but followed a winding or an oblique
course. This is probably because Sennacherib's engineers sought to
utilise the best foundations that they could find.

The dam at Marib in the Yemen, although it was not the only
dam in ancient southern Arabia, is the best known and the most
impressive. Three dams, all on the same alignment, were built
across the Wadi Dhana in succession, in order to impound flood
waters to irrigate the lands around the city of Marib. The first of
these was built by the Sabaeans about 750 BC. No trace of it
remains, but it was built of earth and was about 2,000 feet long and
about 13 feet high. The second dam was built about 500 BC in
order to allow the area under irrigation to be increased. Some
traces of this remain: it was also an earthen bank about 2,000 feet
long, but the height was increased to 23 feet. The third dam was
built by the Himyarites, after they had succeeded the Sabaeans as
rulers in 115 BC. In this construction the height was doubled to 46
feet and more elaborate waterworks were installed. These consisted
of a five-channel spillway at the northern end, and a feeder channel
which led into a settling tank from which it flowed through a paved
channel, 3280 feet long, into a distribution tank from which it ran
into 14 separate canals. For this dam, the southern sluices, originally built for the second dam, were rebuilt and extended. To

Figure 3.1: Hydraulic Works at Marib

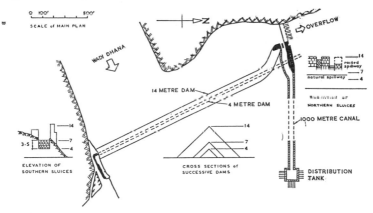

The Marib dam: the successive dams built on the site are shown in plan and cross-
section. The figure also shows elevations of the sluices.

support the end of the earth dam an artificial masonry abutment was built up, and between this wall and the rock face ran the outlet channel. This was the same channel that the second dam supplied, but now its floor was raised $11\frac{1}{2}$ feet and masonry walls were constructed to contain the flow. The southern sluices in their final form featured masonry construction of very high quality, the carefully cut and fitted blocks using cast lead dowels in their joints but no mortar (see Figure 3.1).

The Nabataeans were a people who flourished in southern Jordan and in the Negev from the first century BC to the end of the first century AD, when they were overrun by the Roman armies of Trajan. They developed very elaborate hydraulic systems for utilising the runoff water from occasional showers. For this purpose they built many small dams across wadis in order to divert the water to the fields, either directly or through storage tanks. In the area around the ancient city of Ovdat in the Negev the extent of these waterworks is astonishing: some 17,000 dams have been located in an area of 50 square miles. The dams built to divert water into canals were generally the largest and were built of masonry. Smaller dams were built to conserve water and silt in the wadis themselves. In these the construction was usually composite — a rubble and earth core faced with roughly shaped masonry blocks. Their average length was about 150 feet, their average thickness 7 feet, and their height about 6 feet.

The Romans

The Romans learnt much of their civil engineering from other peoples. They probably inherited the arch from the Etruscans, from the Greeks they learned about water supplies by means of aqueducts, techniques of tunnelling and the use of hydraulic cement, and in the Negev they took over and maintained the Nabataeans' irrigation systems, no doubt with the assistance of Nabataean engineers. The Romans have rightly been called the greatest civil engineers of ancient times, but the key to their success lay in their ability to organise and administer large-scale projects, and in their practical approach to engineering problems. They did not, however, contribute very many fundamentally new ideas or original concepts. Their achievements were remarkable, and the sheer size and scope of their constructions, and the will and determination that brought them into being, excite our admiration.

Dams were built in all parts of the Roman Empire. Major constructions include three in Italy, one in France, five in the Iberian peninsula, seven in North Africa, three in Asia Minor, and six in Syria. It is somewhat surprising that only three dams have been located in Italy itself, all of them near the town of Subiaco, about 50 miles east of Rome. All of them were built in the middle of the first century by Nero's engineers to create artificial lakes for his villa by damming the river Aniene. This in itself is unusual — the construction of dams for recreational rather than utilitarian reasons. A second point of interest is that the dams, which were built of masonry, were very large. One of them was at least 130 feet high, easily the highest dam which had been built anywhere in the world up to that date. Nothing comparable was to be attempted until 1,500 years later.

The town of Glanum, based on an earlier Greek site in Provence, was a very early Roman settlement founded in the second century BC. It is not known precisely when the town was provided with a water supply, which was in the form of an aqueduct fed from a small curved dam. It is possible that this was a true arch dam, although the dam built on the same site in 1891 is an arched gravity dam, because it resists the water pressure primarily by its weight. As far as can be ascertained from pre-1891 sources, the Roman dam was made of two masonry walls, each a little over 1 metre thick with a space of about $1\frac{1}{2}$ metres between them that was probably originally filled with earth and stones. It was about $3\frac{1}{2}$ metres thick by 6 metres high. These dimensions, and the location of the dam in a deep, narrow gorge, indicate that it may well have acted as a true arch dam. If this is the case, it is one of the very few examples of this type built before modern times.

Two of the Roman dams in Spain are still in use today. Both of these were built to supply the town of Merida with water. The Proserpina dam, probably built in the first third of the second century AD, was 1,400 feet long with a maximum height of 40 feet. It was built of masonry with a core of concrete, but the thickness at the crest is only 7–8 feet in places, and it was therefore necessary to buttress the downstream face with an earth embankment. Outlet was through two wells lined with masonry built into the structure of the dam and connected to the reservoir by tunnels, presumably controlled by sluices of some kind (nowadays large-diameter iron pipes with screw valves are used). From the bottom of the wells, the water was discharged through conduits running through the earth bank. The Cornalvo dam of somewhat later date, is 66 feet high and 656 feet long (Plate 1). Its internal structure consists of sets of intersecting walls, longitudinal and transverse, which divide the interior into a series of deep masonry boxes. These were filled either with stones or clay and the whole arrangement was then covered with earth. On the upstream side a masonry facing was provided to stop the water washing the soil away.

Many dams were built by the Romans in North Africa, and there are marked similarities between the dam-based systems in North Africa and those in the Negev. This is hardly surprising when we remember that the Romans took over and maintained the Nabataean systems; it was therefore natural for them to apply Nabataean techniques in Libya and Tunisia, where the climate and topography were similar to those of the Negev. Notable Roman dams in the region are known at Kasserine in Tunisia and Leptis Magna in Libya. The dam at Kasserine was about 500 feet long by 33 feet high, with a trapezoidal cross-section 24 feet at the base and 16 feet at the crest. It was built of cut and fitted masonry blocks with mortared joints and a rubble-and-earth core. There is a very important group of dams at Leptis Magna. Most of these were for the usual purposes of water supply and irrigation, but one of them was built to protect a water supply reservoir from the muddy water which came rushing down after a rainstorm. At 2,950 feet, this is one of the longest dams ever built by the Romans.

A very large wadi-irrigation dam in Syria is still intact, at Harbaqa, about 45 miles south-east of Homs near the ruins of a castle called Qasr al-Hayr al-Gharbi. It is made of dressed masonry with a concrete core and has a length of 365 metres and a height of 21 metres; a motor vehicle can be driven from one side to the other. The upstream side is filled with silt almost up to the crest. About eight miles south-west of Homs is the impressive Roman dam over the River Orontes, which forms the huge reservoir known as the Lake of Homs. The dam was repaired and raised in 1934, but examinations carried out before that date have preserved the most important details of the original construction. It was built in 284 AD by the Emperor Diocletian and formed a reservoir 6 miles wide by $2\frac{1}{2}$ wide that was used for both water supply and irrigation. The core was made of the usual mass of rubble masonry, bound together with a whitish hydraulic mortar, very hard and strong, faced with dressed basalt blocks. The dam is nearly 2 kilometres

long with a maximum height of 20 feet, at which point the crest is
23 feet wide and the base perhaps 66 feet. Along the structure at
various points were sluices, designed to run water into canals. The
dams in Turkey are not of the usual Roman construction since they
are built with vertical walls. The dam at Örükaya in north-west
Anatolia, for example, has vertical walls of dressed stone, dowelled
together with lead; the gap between them was filled with rammed
earth. The total width of the dam is 5 metres and its height 16
metres, and it was supported on the downstream side by
buttresses.[3]

Byzantium and Iran

The water supply of Constantinople goes back to the time of
Hadrian, but it is not known for certain what the works consisted of
in that period. Four of the eight dams that were built later to
provide the city with water are, however, still in use, all of them
constructed over two small rivers that feed the Golden Horn. The
largest of these forms the Buyuk Bent or Great Reservoir. It has a
total length of 250 feet, a maximum height of 41 feet and the
thickness of the base of its trapezoidal cross-section is 33 feet. It is
constructed of a rubble masronry core faced with roughly squared
masonry blocks. It is now provided with spillways but it is unlikely
that these were part of the original installation, in which surplus
water probably flowed over the crest of the dam. Openings in the
dam's face connected with an internal passage that conducted water
to the masonry aqueduct leading to Constantinople, about nine
miles away. How these outlets were controlled is not known. The
date of this dam, and of the other three serving Constantinople is
not known, but the historian Procopius has left us a record of the
construction of a dam at Dara in Upper Mesopotamia in the first
half of the sixth century. The dam, which was built in the reign of
Justinian and may have been commissioned by him, was built just
outside the walls of Dara to control the floods of the river that
flowed past the town. It had two sluices at different levels, one a
low-level outlet, the other probably a high-level spillway. The dam
was curved and Procopius says 'the curve, by lying against the
current, might be able to offer still more resistance'. We have a
clear indication here that this was a true arch dam, the second that
we have encountered.

There is a long history of dam building in Iran. The capital of the Achaemenids in the sixth century BC was Persepolis and a number of dams were built over the river Kur, which flows past the city, to irrigate the surrounding lands. There is no doubt that some of the ancient dams still standing on the Kur are Achaemenid in origin. The dynasties which ruled in Iran after the conquests of Alexander the Great, late in the fourth century BC, did not contribute much to engineering, but after Ardashir became the first Sasānid ruler of Iran in AD 226, there was a marked increase in constructional activity. Wars were numerous, however, and in one of these in about AD 259 the Roman Emperor Valerian, together with his army of 70,000 men were captured by the Persian army under Shapur I. The prisoners were taken to Shushtar and given the job of damming the river Karun. The first job was to dig an artificial cut, the Ab-i-Gargar, to divert the entire flow of the Karun. When the bed of the river below the diversion had dried out, a huge dam, which still exists, was constructed. It is 1,700 feet long and has a rubble masonry core set in hydraulic mortar, and the facing is of large cut masonry blocks held in place with both mortar and iron clamps set in lead. The dam is pierced with numerous sluices for the purpose of releasing water in times of excessive flow. A bridge, the Pul-i-Kaisar, also existing, was built on top of the dam. When the work was complete the entrance to the Ab-i-Gargar was closed with a second dam, and this too has survived. There can be little doubt that Valerian's army contained engineers, and that they therefore made a significant, if forced, contribution to the development of hydraulic engineering in Iran. The Romans' activities at Shushtar are particularly important because for the first time we have evidence of a river being totally re-routed so that a dam could be built in the dry. As mentioned earlier, it is difficult to see how a dam could be built over a major river without a diversion of this kind.

The success of the Shushtar dams led to the construction of others. One of these was at Ahwaz, and it was over 3,000 feet long and, as far as can be judged, about 25 feet thick. Al-Muqaddasī, writing towards the end of the tenth century, describes it as a masonry dam, with wonderful fountains. The reservoir which it formed came almost up to the level of the city, and below the dam the water was conducted into three canals from which the arable land was cultivated. He continues:

They say that were it not for the dam Ahwaz would not be occupied and its rivers would be useless. In the dam are gates which are opened if the water is abundant so that Ahwaz is not flooded. The noise of the descending water prevents sleep for most of the year.[4]

The Muslims

Nearly all the dams described in the previous sections were under Muslim control by the middle of the eighth century, and the Muslims could therefore draw upon a number of sources as an inspiration for their dam-building undertaking. The example of Roman dams was especially prominent because of the scale on which they worked and because their influence was relatively recent. Contrary to the statement of some historians there was no decline in engineering activities, including dam building, in Muslim times. On the contrary, the new irrigation systems and the extensions to existing ones described in the previous chapter necessitated the construction of a large number of dams. Many of these were small diversion dams, but several large dams were built in Iraq, for example, including three across the Tigris. The finest dam in the neighbourhood of Baghdad, however, was on the Uzaym river. This was built of cut masonry blocks throughout, connected with lead dowels poured into grooves — a fairly common Muslim technique. The dam was 575 feet long, with a trapezoidal cross-section, 10 feet thick at the crest and 50 thick at the base; its maximum height was something over 50 feet, but this rapidly reduced towards the sides.

The Muslims built a number of dams in Iran, including the addition of one, the Pul-i-Bulaiti, to the Shushtar system. One of the great builders of the Buwayid dynasty, who held the real power in Iraq and Iran from 945 until 1055, was 'Adud al-Dawla. His most impressive dam was the Band-i-Amir, built in 960 over the river Kur in the province of Fars between Shiraz and Istakhr (the ancient Persepolis). To quote al-Muqaddasī again:

'Adud al-Dawla closed the river between Shiraz and Istakhr by a great wall, strengthened with lead. And the water behind it rose and formed a lake. Upon it on the two sides were ten water-wheels, like those we mentioned in Khuzistan, and below each wheel was a mill, and it is today one of the wonders of Fars. Then he built a city. The water flowed through channels and irrigated 300 villages.[5]

The dam, which still survives, is built of solid masonry, set in mortar and reinforced with lead dowels. It is some 30 feet high and 250 feet long.

Many dams were built by the Muslims in southern Spain, mainly for irrigation but also for water supply and water power. There was, for example, a series of dams in the region of Valencia (Plate 2). At Murcia (Plate 3), a series of low dams, like those of Valencia, was not a practicable solution for hydrological and topographical reasons, and a single, carefully sited dam was therefore built. This was 420 feet long and 25 feet high. For three-quarters of its length the width of its base is 160 feet, reduced to 125 feet for the other quarter. The reason for this apparently excessive width is that the bed of the river is very soft and weak. It is supposed that the Muslims built a massive dam, probably carried on a bed of piles, to give the structure enough weight to resist the water pressure and to ensure that it was not able to slide along the river bed. That the dam has successfully withstood 1,000 years of use seems to indicate that uplift, caused by water percolating under the foundations, has not been a problem. There is another interesting feature of the dam, which was built of rubble masonry and mortar faced with large masonry blocks. In Norman Smith's words:

> The large surface area of the dam's air face was put to good use. Water flowing over the crest initially fell vertically through a height of 13–17 feet on to a level platform, 26 feet wide, running the length of the dam. This served to dissipate the energy of the water spilling over the crest. The overflow then ran to the foot of the dam over flat or gently sloping sections of the face. In this way the whole dam acted as a spillway and, moreover, the energy gained by the water in falling 25 feet was dissipated *en route*. Thus the risk of undermining the down-stream foundations was greatly reduced. It is for reasons like these that one is inclined to believe that the Moslems had some understanding, albeit empirical, of hydraulics.[6]

Europe

By the middle of the thirteenth century the whole of Spain except
for the province of Granada in the south had been reconquered by
the Christians. The Muslims were to remain in Granada for another
250 years, but there were also communities of Muslims in many
parts of Christian Spain. In certain fields, including agriculture and
technology, Muslim influence remained strong, and it is therefore
hardly surprising to find that the first dams built by the Christians
showed no radical departure from Muslim practices. There was one
important innovation, however, or rather reintroduction: the
Roman system of building dams to form permanent reservoirs was
followed by the Spaniards in the thirteenth and later centuries.
Notable among dams built for this purpose are the Almonacid de
Cuba dam, about 25 miles south-east of Saragossa, and the
Almansa dam three miles to the west of the town of that name. The
first was built in the first half of the thirteenth century and it is
massive — 340 feet long, nearly 100 feet high and nearly as thick.
The core is of rough rubble masonry embedded in a mixture of
stones, earth and lime mortar. The air face consists of carefully cut
blocks of masonry arranged in steep steps (Plate 4). The Almansa
dam, probably built early in the fourteenth century, is of similar
construction but smaller. It is noteworthy for being curved,
although it is heavy enough to act as a gravity dam (Plate 5).

In 1179 work was begun on an irrigation canal in northern Italy
from a point on the River Ticino near Oleggio to Milan. The canal,
known as the Naviglio Grande, has been fed from substantially the
same dam for nearly 800 years. A number of reconstructions have
been necessary over the centuries, and it is not therefore possible to
know what the original dam was like. Its present dimensions are
920 feet long, about 6 feet long and varying in thickness from 60
feet down to 30 feet; the medieval dimensions may have been
similar. Another dam was built in the Milan area in the thirteenth
century and a further five in the fourteenth. Apart from these dams,
the oldest dam in Italy, built about 1450, stands over the River
Savio near Cesena. The dam is still in service, and despite some
modifications, the profile of the fifteenth century dam can be
deduced. Its vertical water face was some 19 feet high and the crest
a few feet wide. Below the crest, the air face sloped downwards for
a length of more than 40 feet and then dropped about 5 feet to the
level of the foundations. The dam, which is 234 feet long and about

45 feet thick at the base, is built of brick and timber. Although these materials are not ideal for a river dam without proper spillways, the dam was obviously a success.

By the twelfth century at the latest, large dams were being built in northern Europe to increase the power delivered to water-mills. One such undertaking was in the Toulouse area, where three dams were built over the River Garonne; together they supplied power to 43 mills. One of these, the Bazacle dam, first mentioned in a document of 1177, was 1,300 feet long and built diagonally across the river. Like the other dams it was built by ramming thousands of oak piles about six metres long into the river bed. A series of parallel palisades was thus formed, and the spaces between were filled with earth, wood, gravel and boulders. Breakwaters were built in front of the dam to protect it from floating debris.[7]

Notes

1. Norman Smith, *A History of Dams* (Peter Davies, London, 1971), pp. 1-5. Most of the remainder of this chapter is based directly on Dr Smith's work. Many of the descriptions in his book are derived from his own examinations of surviving dams, or from the published descriptions of other eyewitnesses.

2. Raf van Laere, 'Techniques Hydrauliques en Mésopotamie Ancienne', *Orientalia Lovaniensa Periodica*, University of Leuven, Belgium, 11 (1980), pp. 11-53, pp. 49-50.

3. Niklaus Schnitter, 'Romische Talsperren', *Antike Welt*, vol. 2 (1978), pp. 25-32, p. 31.

4. Al-Muqaddasī, *Ahsān al-taqāsīm fī ma'rifat al-aqālīm*, ed. M.J. de Goeje in BGA vol. 3 (Brill, Leiden, 1906), pp. 411-12.

5. Ibid., p. 444.

6. Smith, *History of Dams*, pp. 95-6.

7. Jean Gimpel, *The Medieval Machine* (Victor Gollancz, London, 1977), pp. 17-18.

4   Bridges

The origin of cantilever (Figure 4.1) and suspension bridges almost certainly lay to the east of our area. In the south and west of China, in the Himalayan regions and in Afghanistan, cantilever bridges had been built from, at latest, the fifth century AD. Simple cantilever bridges were built in Savoy throughout the Middle Ages, and several sketches of them appear in the notebooks of Villard de Honnecourt (*c.* 1235), but there is no record of their use in western Asia or Europe in earlier times. Being made of timber, they do not have a long life nor do they leave traces, particularly since in many cases modern bridges may have been built at their sites. It would be surprising, however, if they were not built in mountainous areas such as the Zagros and the Taurus, since they are an excellent method of crossing deep ravines in hilly country. Usually a timber substructure is built into a masonry abutment on either bank, and the longitudinal and transverse beams to carry the roadway extend from the top of this supporting structure. At the centre the two cantilever spans support a short beam section. There is no need to build piers and no need to descend to the bottom of the ravine. In the tenth century a bridge over the River Tāb in Iran was described briefly by Ibn Hawqal. He says that the river was crossed by a wooden bridge 'suspended between the sky and the water, its height above the water about 10 cubits'.[1] He may, of course, have seen a suspension bridge, but the cantilever type seems more likely, especially since he makes no mention of ropes. The evidence is tenuous, but it seems a reasonable conjecture that this method of construction was transmitted from East Asia to western Europe via Iran, Anatolia and the Balkans.[2]

Suspension bridges with cables made of woven bamboo strips were used in China, perhaps as early as the third century BC, and certainly by the first century AD. Their use was widespread in China, Tibet, Afghanistan, Kashmir, Nepal, India, Assam, Burma and Thailand. Indeed, communication between peoples in many parts of those lands would have been almost impossible without the sus-

Figure 4.1: Bridge Types

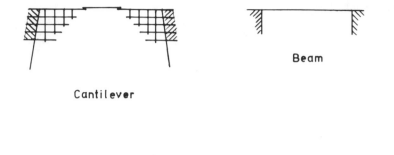

Cantilever

Beam

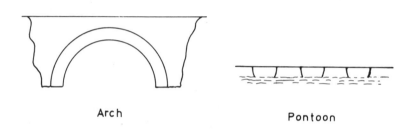

Arch

Pontoon

pension bridge. The iron-chain suspension bridge was introduced to China about the sixth century AD.[3] It seems strange, therefore, that we have no record of the technique in western Islam or Europe before the publication of Faustus Verantius' book on engineering in 1595, which contains an illustration of a fairly advanced iron-chain bridge together with a brief description.[4] The Muslims had conquered most of Afghanistan by the end of the seventh century, and entered into trading relationships with China and India from the eight century onwards. They must have seen suspension bridges and they were able to make both heavy cordage and iron chains. But if suspension bridges were used in western Islam there is no record of the fact.

On Trajan's column in Rome there is a representation of the bridge that he had built over the Danube in AD 104 by Apollodorus of Damascus. This was probably a flat arch truss, similar to the Alconetar bridge that he built in Spain. The first sure evidence for the use of trusses in bridging in western Europe is the bridge built by Palladio (1518-80) over the River Cismone, a tributary of the Brenta. This was of timber construction, but in 1595 Verantius set out in detail the use of metal in bridge trusses. Another type of bridge, which is of relatively minor importance, is the drawbridge. This is familiar to everyone as the means of crossing moats that surrounded medieval fortresses; ropes were attached to the end of the bridge that was on the outer bank of the moat and these passed over pulleys inside the castle, where they passed over the drum of a winch; the other end of the bridge was hinged. The bridge was raised at nightfall or when an enemy approached. Such bridges were also used for peaceful purposes. In the first half of the fourteenth century there was a wooden bridge at the town of Ashmun on the lower Nile. Once a day part of the bridge was raised to allow ships to pass through.[5] Towards the close of the fifteenth century the Rialto bridge in Venice was rebuilt in timber, with a central section that could be raised to allow vessels with masts and sails through.[6]

Only three types of bridge were important in the classical and medieval periods in western Asia and Europe: beam, pontoon and arch bridges. These can be dealt with most conveniently by devoting a separate section to each.

Beam Bridges

The simple beam bridge in the form of a log or a flat slab of stone placed over a stream has its origin in prehistoric times. It has the advantage of simplicity, but its applications were limited before the introduction of modern materials such as structural steel and reinforced concrete. The stresses that are produced when a beam is loaded are shear stresses, at a maximum at the abutments, and bending stresses, at a maximum at mid-span. In a beam of uniform cross-section, the stress due to bending is compressive in the top half of the beam and tensile in the lower half. It is the tensile stress that causes the problems, because of the three basic building materials available in early times — masonry, concrete and timber — only timber can withstand significant tensile stresses. It is hardly surprising, therefore, that nearly all early beam bridges were constructed of timber, but even with this material there are serious disadvantages. Timber is much weaker than steel, and simple beam bridges are limited to a practicable span of about 20 feet. It is possible to increase the span by fitting struts or a timber crib at the abutments, so reducing the unsupported span. A further disadvantage is that timber is liable to damage by storms, floods or fire, and in any case has a fairly short life. This is also unfortunate for the historian, since no substantial parts of old timber bridges have survived, and reliance has therefore to be placed on written and iconographic evidence. This evidence is in the form of scattered references in historical and geographical works.

The Romans built their first bridges of timber, and this form was retained for military uses and for most of the bridges in the outlying provinces. The earliest recorded Roman bridge was the Pons Sublicus, built over the Tiber in Rome about the sixth century BC. A timber bridge, it is famous as having been held by Horatius and his two comrades against the whole Etruscan army, while the structure was destroyed behind them. Another well-known example is the wooden trestle bridge built by Caesar's engineers over the Rhine in 55 BC. Piles were driven into the river bed at an angle, and these were strengthened at the sides by raked supports. Transverse beams were fixed across the top of the piles to form trestles, upon which the longitudinal beams and the timber decking were placed. The bridge was protected against floating timber and other debris by palisades of piles driven upstream. The entire work was completed in ten days. During dredging work on the River Trent in England in 1885, the remains of a Roman bridge were found. There were seven stone piers, spaced 29 feet apart, each pier provided with slots to carry a wooden superstructure.[7] Wooden bridges over the River Tāb in Iran, on the Nile and in Venice have already been mentioned. References to this kind of bridge are rare, but there is no doubt that simple wooden bridges were very common up to the end of the Middle Ages and beyond. Their total contribution to the needs of communal life and commerce was probably no less than that of the more spectacular (and more durable) masonry bridges.

Pontoon Bridges

Historians of engineering usually devote only a few lines to pontoon bridges — bridges of boats — before getting down to the serious business of discussing arch bridges. Yet this type of bridge was an important and widely used method of crossing rivers in the classical and medieval period, especially in the Muslim world. Again, we are dependent mainly upon literary sources for our information, but there are more references to this type of bridge than is the case for beam bridges. The idea may have originated when boats were moored side by side at a jetty or landing stage, and planks were placed across the gunwales of two adjacent boats to make it easier to walk from one boat to the other. This type of bridge can be constructed quickly, and it is still the only type of bridge suitable for crossing rivers more than about 150 feet wide during military operations. It is quite simple to build a pontoon bridge in still water, but the skills of watermanship are needed when a current is flowing, to manoeuvre the boats into position and anchor them at the correct distance apart, ready to receive the beams and decking for the roadway. Anchors are not sufficient to hold a pontoon bridge in position for long periods on a swiftly flowing river, and it is therefore necessary to stretch a heavy cable across the river upstream from the bridge and secure it to anchorage on each bank. For added security, another cable may be laid on the downstream side. Each boat is made fast to the cables. The disadvantages of pontoon bridges are that they require constant maintenance, that they are liable to damage by floating debris or floods, and that they form an obstacle for river traffic.

The earliest record of a bridge used to cross a river is the evidence of an embossed gold band in the British Museum showing the Assyrians using pontoon bridges in the ninth century BC. One of the fullest descriptions of such a bridge occurs in the writings of Herodotus, and refers to the crossing of the Hellespont by Xerxes in 480 BC. The following translation is by Aubrey de Selincourt (Penguin Classics, 1954):

Galleys and triremes were lashed together to support the bridges — 360 vessels for the one on the Black Sea side, and 314 for the other. They were moored head-on to the current — and consequently at right angles to the actual bridges they supported — in order to lessen the strain on the cables. Specially heavy anchors were laid out both upstream and downstream — those to the eastward to hold the vessels against winds blowing down the straits from the direction of the Black Sea, those on the other side, to the westward and towards the Aegean, to take the strain when it blew from the west and south. Gaps were left in three places to allow any boats that might wish to do so to pass in or out of the Black Sea.

Once the vessels were in position, the cables were hauled taut by wooden winches ashore. This time the two sorts of cable were not used separately for each bridge, but both bridges had two flax cables and four papyrus ones. The flax and papyrus cables were of the same thickness and quality, but the flax was the heavier — half a fathom of it weighed 114 pounds. The next operation was to cut planks equal in length to the width of the floats, lay them edge to edge over the taut cables and then bind them together on their upper surface. That done, brushwood was put on top and spread evenly, with a layer of soil, trodden hard, over all. Finally a paling was constructed along each side, high enough to prevent horses and mules from seeing over and taking fright at the water.[8]

As the text indicates, two bridges were used for the crossing, and they were the second attempt — the bridges built earlier in the same campaign had been washed away by a violent storm. The arrangement of the cables was unusual by later standards, since they were used not only for additional security, but also as the foundation for the roadway. We are not told how the bridge traffic crossed the gaps left for the ships, but we must assume that there was some means of opening the bridge temporarily to let ships through. Otherwise the account is straightforward and describes a sound piece of military engineering. The brushwood and earth cover was necessary to protect the wooden decking from the pounding of the animals' hooves. Without it the timber would very quickly have been cut to pieces.

There are frequent references to pontoon bridges in the works of Arabic writers. They were very common in Iraq, for crossing the two rivers and the major irrigation canals. In the tenth century there were two bridges over the Tigris at Baghdad, but only one was in use; the other, having fallen into disrepair, was closed, because few people used it.[9] Ibn Jubayr, writing towards the close of the twelfth century, describes a bridge of large boats over the Euphrates at Hilla. It had chains on either side 'like twisted rods' which were secured to wooden anchorages on the banks.[10] He also mentions a similar, but larger, bridge over a canal near Baghdad.[11] There were also pontoon bridges on the rivers of Khuzistan, the province adjoining Iraq,[12] and on the Helmand river in Sijistan (now western Afghanistan).[13] There seems to have been a pontoon bridge at Fustat (now Old Cairo) in Egypt for many years. In the first half of the tenth century, al-Istakhrī says that one bridge crossed from the city to the island and a second bridge from the island to the far bank of the river.[14] About two centuries later, al-Idrīsī describes the same arrangement, adding that there were 30 boats in the first bridge and 60 in the second.[15] Bridges of boats do not seem to have been so common in medieval Europe, but some were built. The first Rialto bridge in Venice, for example, was a pontoon bridge built in the thirteenth century.[16]

Figure 4.2: Types of Arch

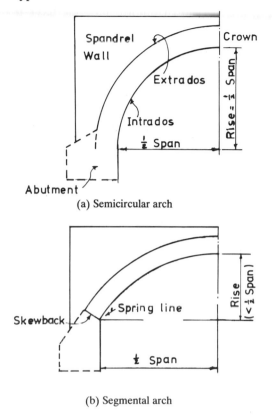

(a) Semicircular arch

(b) Segmental arch

Arch Bridges

The two main types of arch are shown in Figure 4.2: (a) is the semi-circular arch, (b) the segmental; the terms used in designating the parts of masonry arches are shown on (a), but these apply to all types of arch. Apart from segmental arches, i.e. those in the shape of the chord of a circle, other curves are used, including elliptical, parabolic and multi-centred curves, but these need not concern us here as they were not introduced before the Renaissance. Before modern times all arches were fixed, that is, there were no hinges at the abutments or the crown. The mathematical analysis of arches, quite a complex matter, was not developed until the nineteenth century; before then the engineers relied upon empirical data derived from past practice. It is worthwhile, however, to mention briefly the basic points of arch behaviour. At piers and abutments the reactions from an arch consist of two components, a vertical thrust and a horizontal thrust. Where two arches meet at a pier, the horizontal thrusts will cancel each other out, but the abutments must be able to withstand both components. Each section of the arch is subjected to a shearing stress and (possibly) a bending couple. The bending moment on the arch rib is reduced to zero wherever the thrust line passes through the neutral axis. It is therefore highly desirable so to select the arch ring curve that the thrust line to as great an extent as possible fulfils this requirement. Bending produces tensile stresses and, as we have seen, masonry and concrete are weak in tension. The early engineers could not, of course, have expressed the situation in these terms, but they were aware of the need for good workmanship in the arch ribs and for having adequate thickness, especially at the haunches. The imposed stresses, whether due to shear or to bending, are most severe at the quarter-span, the point where ruptures most frequently occur. Much more frequent, however, are failures of the foundations, a matter to which we will return a little later.

It is often a matter for comment that a number of old arch bridges that were intended to carry only pedestrians, pack animals and carts, are able to withstand the full load of modern motor traffic. The inference is therefore drawn that the bridges were overdesigned, but this kind of thinking is an anachronism, and is also unjust. When an arch is constructed with solid spandrels, with firm foundations and superstructure, the greatest part of the imposed stresses comes from the weight of the structure itself, and in the days before motor traffic the stresses due to the traffic on the bridge were negligible. In other words, if the bridge could carry its own weight, the passage of people and animals over it would have made no noticeable difference. And even with modern traffic the stresses due to the live load are less than those due to the weight of the structure.

In a semicircular arch the ratio of rise to half-span is 1.0, because the rise is equal to the radius of the curve. If the ratio is less than unity, the curve is segmental, and the smaller the ratio becomes the more daring is the design. The bridge is more elegant, lighter and can be built with a longer span, thus reducing the number of piers required. In China the first known segmental bridge was built by the engineer Li Chhun in AD 610, with a rise/half-span ratio of .38 and a span of 123 feet,[17] and others were built in that century and later. Bridges with this degree of segmentality were not constructed in Europe until the fourteenth century, although the Pons Fabricius in Rome (62 BC) and the bridge at Avignon (AD 1177) both have a rise/half-span ratio of .83. It is quite possible that this type of arch was transmitted to Europe in the Middle Ages by the reports of travellers, but equally possible that it was invented independently when the need arose. When it was necessary to limit the number of piers, or to have none at all, while at the same time restricting the height of the arch, the segmental arch almost chose itself. The reasons why the Romans did not adopt the segmental arch may not all have been due to conservatism: in general they had better foundation strata than the Chinese, and their use of hydraulic cement, which was not known in China, reduced the vulnerability of their structures. It is worthy of note that the spans of a number of Roman semicircular arches were equal to or longer than the spans of Chinese segmental arches.

It has now been established that the arch, the vault and the dome were known to Sumerian Mesopotamia, a fact that may explain the appearance of these forms in China and the West some time in the first millenium BC. The Greeks did not use arches to any great extent, whereas arched structures are almost a hallmark of Roman civilisation. Indeed, so many Roman bridges still survive, that even to list them would occupy all our available space. The second century BC witnessed the start of an unprecedented activity in the construction of masonry arch bridges. The earliest bridge over the Tiber, the Pons Aemilius, was built in 179 BC, originally with a timber superstructure, which was converted to arched construction in 145 BC. This was followed by a further seven bridges. Of the original eight, six are still standing, most of them more or less rebuilt and repaired. The Pons Fabricius, however, is in an excellent state of preservation and is an impressive structure with spans of about 80 feet. In the provinces many Roman bridges still stand, notably the bridge at Pont Saint-Martin, between Ivrea and Aosta, with a single span of 117 feet, and the magnificent six-arch bridge over the Tagus in Spain, built under Trajan; its two central arches each have a span of about 87 feet.[18] The many arched bridge over the Guadalquivir at Cordoba was described by al-Idrisi,[19] and indeed it is noteworthy that Muslim writers often refer to Roman bridges with admiration. One in particular that impressed all the Muslim writers who saw it is the bridge at Sanja in Upper Mesopotamia. This was built by the Emperor Vespasian about 70 AD and it is a single arch of dressed masonry with a span of 112 feet.[20]

The first of the Roman aqueduct bridges was the Aqua Marcia, built in 144 BC. Some of its arches still survive — they have a span of 18 feet. Impressive remains of many aqueduct bridges can be seen in Italy, France, Spain and North Africa (Plates 6, 7). Among these some of the most impressive are the Pont du Gard, near Nîmes in France, built of cut stone blocks and consisting of three tiers of arches with a total height of 160 feet (*c* 20 BC); Tarragona, complete in the form of a two-tiered bridge with only the roof of the conduit missing (first century AD); Merida (first century AD) (Plate 8); the tall, slender bridge over 100 feet high, made of large granite blocks with unmortared joints that brought water to Segovia (time of Claudius); Cherchel in North Africa, a bridge standing 115 feet high (time of Hadrian).[21]

Roman bridges — both for river crossings and aqueducts show a noticeable evolution in structural techniques with the passage of time. The application of concrete was a major factor in this evolution. The Pont du Gard was beautifully constructed of cut masonry blocks, and successfully reached a great height by stacking one row of arches on top of another, but the technique is clumsy and expensive. The Segovia and Tarragona bridges are good examples of the more elegant and structurally economical shapes evolved in the first century AD. Even more refined forms developed as concrete came into wider use. In the Merida and Cherchel aqueduct bridges the 'stacking' form has disappeared — the tall slender piers, constructed of concrete cores with facings of masonry and brick, are continuous from top to bottom.

There was a continuous tradition for the construction of arched bridges in the East, especially from the third century AD onwards. The great single arch bridge built by Justinian over the River Sayhan at Adana was rebuilt by the Muslims in 743 and again in 840.[22] In Sasānid Iran a large number of arch bridges were built, and these begin to show Roman influence after the bridges at Shushtar (Arabic Tustar) had been built by Roman prisoners of war from AD 260 onwards. In earlier times the Iranians had not known the use of coffer-dams and were therefore unable to construct foundations in water. They sited their bridges where there were islands in the rivers that could be used for placing the piers. The first bridge in which Roman methods were used for the foundations was at Dizful in Khuzistan. This was built in the fourth century and had an overall length of 1,250 feet.[23] Other notable Sasānid bridges were a bridge of kiln-burnt bricks over the Dujayl at Ahwaz, and a bridge over the river Tāb in central Iran, the remains of which still exist.[24]

Many arch bridges were built in Muslim times, not only over rivers, but also over irrigation canals. Sometimes these were built on the sites of ruined Sasānid bridges. Al-Qazwīnī has left us a graphic description of the rebuilding of a great arch bridge at the town of Idhaj in Khuzistan. It was built by the Wazir of the Buwayid Amir al-Hasan (d. 977) who conscripted craftsmen from Īdhaj and Isfahan. The bridge was 150 cubits in height and consisted of a single arch, supported on tapering masonry piers strengthened with lead dowels and iron clamps. The slag from iron workings was used to fill the space between the arch and the roadway.[25] Another remarkable bridge, not far from the Sasānid bridge over the River Tab, is described by al-Istakhrī. He says that it was built by an Iranian, physician to the Umayyad governor al-Hajjāj (661-714). It was a single arch of span about 80 paces, and so high that a man on a camel with a long standard in his raised hand could pass beneath it.[26] A bridge with an unusual purpose is among the works of Ibn Tulūn, governor of Egypt from 868 to 884. It was, says Ibn Jubayr, a bridge of 40 arches 'as large as the arches of a bridge can be' and it was part of a causeway six miles long that led out from the city of Fustāt in the direction of Alexandria. It was used as a passage for troops over the Nile floods, if an enemy approached from the east. There were many fine arched bridges over canals,[27] but these were not simply links in the road systems. As Thomas Glick comments: 'To the irrigator the primary purpose of bridges was to protect the sides of the canals against fording men and animals. The convenience of the traveller was only a secondary consideration.'[28]

In Europe no bridges of note were built between the end of the Roman Empire and the twelfth century. After that time, however, the building of masonry bridges began again in earnest. One of the earliest of these was the famous bridge at Avignon, begun in 1177; part of it still stands today. Each span was over 100 feet, the piers were 25 feet thick and its total length was nearly 3,000 feet. As mentioned earlier, the arches of this bridge were not quite semi-circular, having a rise/half-span ratio of .83. Old London Bridge was completed in 1209 and had a total length of 926 feet and a width of 26 feet. It was carried on 20 arches of roughly semicircular form, supported on 19 artificial islands, which so restricted the channel that during the rise and fall of the tides the river rushed through in the form of rapids. As was the case with many medieval bridges, London Bridge carried a large number of houses, which were destroyed by fire and rebuilt three times before they were finally demolished in 1754.

The bridge itself was finally pulled down in 1832 when John Rennie's bridge was built.

During the fourteenth century a number of magnificent bridges were built with arches of long span and a marked degree of segmentality. In Florence the present Ponte Vecchio was built in 1367, after its predecessor had been carried away in a flood; the longest arch has a span of 98 feet and a rise of 18.4 feet. The Ponte del Castelvecchio at Verona, destroyed in the Second World War but quickly and beautifully restored, had three spans, the greatest of which was 160 feet long with a rise of 45 feet; it was built in 1354–6. Other bridges with segmental arches were built in Italy and France, but the most daring of all was the bridge over the Adda at Trezzo in Northern Italy, built for Bernabò Visconti 1370-7. It consisted of a single arch 236 feet long with a rise of 70 feet. It served as an approach to the ducal castle and was deliberately destroyed in the course of hostilities in 1417. Today all that is left of the bridge are the abutments and two small overhangiing remnants of the arch. It was not until the second half of the nineteenth century that similar spans were attained.[29]

A Note on Foundations

To provide firm foundations for structures is one of the most difficult tasks in the whole field of civil engineering and building. Even today, after the introduction of the techniques of soil mechanics, each foundation is a specific problem and must be considered in relation to local conditions, and the proper interpretation and use of soil test data require both experience and engineering judgement. Both these attributes were needed in large measure by the early engineers. Constructing good foundations on land is hard enough, but in water the difficulties are much greater, firstly because it is difficult to get access to the job site, and secondly because the lower part of the superstructure will be exposed to pressure from the current or the waves, together with scouring action that may attack the boundary layer between the substructure and the supporting stratum. We should not be surprised that some early bridges collapsed, but rather that so many have survived over the centuries.

The Romans' use of hydraulic cement helped in the construction of durable piers for bridges, but is irrelevant to the basic problem of foundations — that of ensuring that the loads are transmitted through the substructure to the supporting material without incurring the risk of settlement. One way of doing this, and it is a good one, is to drive timber piles sufficiently far into the river bed that the frictional resistance of the piles against the soil is greater than the applied load. This method was employed successfully in many early structures, but we do not know how the piles were driven. Vitruvius says simply that it was done 'by machines',[30] but it is a reasonable conjecture that this implies the use of a crane to raise and drop a heavy weight repeatedly. The upper part of the piles could form the legs of a trestle (as in Caesar's bridge over the Rhine), or they could be cut short and capped with masonry or concrete, the pile-caps forming the base for the structure.

It is often desirable to 'de-water' the site before constructing the foundations, since the area can then be excavated down to a good load-bearing stratum. There is no evidence that the diversion of rivers was a common way of achieving this — the usual method of de-watering was to use coffer-dams. Almost any material can be used for the dam, provided that it is reasonably watertight. Once the dam has been built completely around the site, the water is pumped out, and work can start on the foundations. Vitruvius describes the use of this technique, but his description is not entirely clear. It seems that the dam was constructed on shore or on a boat. It was made like a box with a double wall of planks, the space between the two walls filled with clay. It was then floated into position and weights of some kind were loaded on to it until it rested firmly on the bottom of the sea (the passage is describing harbour works). The rest of the description is clear: water was pumped out by means of Archimedean screws or tympanums (see Chapter 8), the foundations were excavated and filled with hydraulic concrete, which was the base for the masonry substructure. If good soil could not be reached wooden piles were first driven over the site.[31]

A similar method is the caisson, which was also known in classical times, although the clearest description of the technique is found in a fifteenth-century work. Mariano Taccola (1382–c. 1458) says that one makes a square box of timber, and then a second wall of planking two feet inside the first. The gap between the two walls is filled with concrete. The caisson 'when half built is floated out to sea using four boats. Each boat has a plug. When the caisson is in position the plugs are removed.'[32] He does not mention how the caisson was finished when in position, or whether the space inside the second wall was de-watered, but the description identifies the main difference between the coffer-dam and the caisson, namely that the former is a temporary structure to enable the foundations to be built in the dry, whereas the latter forms a permanent part of the finished structure.

Notes

1. Ibn Hawqal, *Kitāb Sūrat al-Ard,* Arabic text ed. J.H. Kramers, 2nd edn of vol. 2 of BGA (Brill, Leiden, 1938), p. 249.

2. Joseph Needham, *Science and Civilisation in China* (5 vols. to date, 1954 onwards, Cambridge University Press), vol. 4, pt. 3, 1971, pp. 162-7.

3. Needham, *Science and Civilisation,* vol. 4, pt. 3, pp. 184-210.

4. Faustus Verantius, *Machinae Novae,* ed. Friedrich Klemm (Heinz Moos, Munich, 1965), plate 34, text p. 14.

5. Ibn Battūta, *Rihla,* ed. Karam al-Bustani (Beirut, undated) p. 35.

6. William B. Parsons, *Engineers and Engineering in the Renaissance,* 2nd edn (MIT Press, Cambridge, Mass., 1968), p. 507.

7. Ivan D. Margary, *Roman Roads in Britain* (John Baker, London, 1967), p. 224.

8. Quoted in J.P.M. Pannell, *An Illustrated History of Civil Engineering* (Thames & Hudson, London, 1964), p. 212.

9. Ibn Hawqal, *Kitāb Sūrat,* p. 241.

10. Ibn Jubayr, *Rihla,* Arabic text ed. W. Wright (Leiden 1852); amended version of Wright's text, M.J. de Goeje (Leiden, 1907) p. 213.

11. Ibid., p. 217.

12. At Ahwāz, for example — Istakhrī, *Kitāb al-masālik,* p.62.

13. Istakhrī, *Kitāb al-masālik,* p. 141.

14. Ibid., p. 39.

15. Al-Idrīsī, *Description de l'Afrique et de l'Espagne,* Arabic text with Fr. trans. by R. Dozy and M.J. de Goeje (Brill, Leiden, 1866), pp. 142 Ar./171 Fr.

16. Parsons, *Engineers and Engineering,* p. 89.

17. Needham, *Science and Civilisation,* vol. 4, pt. 3, pp. 175-84.

18. D.S. Robertson, *Greek and Roman Architecture* (Cambridge University Press, 2nd edn, 1943), pp. 236-7.

19. Al-Idrīsī, *Description de l'Afrique,* p. 212 Ar./262 Fr.

20. Ibn Hawqal, *Kitāb Sūrat,* p. 181; Guy Le Strange, *The Lands of the Eastern Caliphate* (Frank Cass, London, 1950; this impression 1966), p. 123.

21. Norman Smith, *Man and Water* (Peter Davies, London, 1976), pp. 87-9.

22. Le Strange, *Lands of the Eastern Caliphate,* p. 131. Brief details of other Byzantine bridges are given by H. Hellenkemper, 'Byzantischer Brückenbau', *Lexikon des Mittelalters* (Artemis Verlag, Munich and Zurich, 1982), vol. 2, pt. 2, pp. 730-1.

23. H. Shirley-Smith, *The World's Great Bridges* (Phoenix, London, 1953), p. 26.

24. Le Strange, *Lands of the Eastern Caliphate,* p. 234, p. 269.

25. Al-Qazwīnī, *Āthār al-bilād wa akhbār al-'ibād,* Beirut, 1960, p. 303.

26. Istakhrī, *Kitāb al-masālik,* p. 91.

27. Ibn Jubayr, *Rihla,* pp. 214-15, for example, mentions the many arch bridges over canals near Hilla in Iraq.

28. Thomas F Glick, *Irrigation and Society in Medieval Valencia* (Harvard University Press, Harvard, 1970), pp. 21-2.

29. Hans Straub, *A History of Civil Engineering,* Eng. trans. E. Rockwell

(Leonard Hill, London, 1952), pp. 48-50; see also Pannell, *Illustrated History*, 214-16; Shirley-Smith, *World's Great Bridges*, p. 33.

30. Vitruvius, *De Architectura* (2 vols., Loeb Classics, Latin text with Eng. trans. by Frank Granger, London 1931; this reprint 1970), vol. 1, bk 3, Ch. 4, p. 181.

31. Ibid., vol. 1, bk 5, Ch. 12, p. 315.

32. *Mariano Taccola and his Book De Ingeneis*, eds. Frank D. Prager and Gustina Scaglia (MIT Press, Cambridge, Mass, 1972), p. 103.

5   Roads

Introduction

The historian of engineering, when he turns his attention to roads, is immediately faced with a problem: of all the societies in western Asia and Europe, from Antiquity until the nineteenth century, only the Romans set out to build a carefully planned road system, with properly metalled and drained surfaces. Many sections of Roman roads still exist and the methods of construction can be determined by properly controlled excavations. Moreover, from a combination of literary and archaeological sources it is possible to build up a fairly comprehensive picture of the road systems throughout the Empire. There is therefore sufficient material to enable books to be written on Roman roads, covering all aspects, including the purely technical. Other societies did not have roads in this sense, or very few, but they did have *communications*. In Islam, for example, where a feeling of cultural identity was fostered by a common religion and the widespread use of Arabic, travel from one province to another — by government officials, merchants, pilgrims and scholars — was a commonplace. Roads were largely unnecessary, partly because of the terrain, and partly because of the universal use of animals, above all the camel, for overland travel and transport of merchandise. They had routes, rather than roads; routes that were recorded and mapped, provided with caravanserais, guard houses, postal stations and water supplies. To concentrate upon the purely technical aspects of roadbuilding, would inevitably leave the impression that only the Romans paid serious attention to communications, an impression that is demonstrably false. In this chapter alone, therefore, I have disregarded my intention of dealing mainly with technical matters, in order to say something about communications in general. For the reasons just given, the discussion of techniques will be found mainly in the section on Roman roads.

Pre-Roman Roads

Iran

The Persian Empires from 550 BC until about AD 640, when Iran was conquered by the Arabs, were not confined to the boundaries of modern Iran, but extended over various regions, depending upon the balance of power between the Persians and their neighbours. Under the Achaemenids Persian rule extended over Iraq, Asia Minor, Syria, Egypt, Macedonia, Thrace and north-western India. The Empire was conquered by Alexander in 334–323 BC, then passed to Seleucus and his successors, who were overrun by the Parthians in the second century BC. The Sasanid Empire was founded in AD 226 by Ardashir, but this Empire, though it witnessed a revival of cultural life in Iran, was not so extensive as the Achaemenid. Apart from a brief occupation of Syria and Egypt in the seventh century AD, its frontier with the Roman Empire, and later the Byzantine Empire, lay more-or-less along the Euphrates. Nevertheless, the routes from Iraq to the Mediterranean remained important for trade — as well as warfare — even when parts of them were not under Persian control. The great Persian route system, under the Achaemenids, led to the palaces of the kings in Susa, Persepolis and Ecbatana. The main route, which Herodotus calls the Royal Road, led from Sardis, the most westerly provincial capital, and from the port of Ephesus, to Susa by way of Issus, Laodicea, the Cilician Gates, Tarsus, Zeugma, Nisibis and Nineveh (see Figure 5.1).[1] This route also gave access to the northern coast of Syria, but in later times it had to compete with the desert routes through the trading towns of Palmyra and Petra. Petra came into prominence in the first century BC, when it was under the control of the Nabataeans, and remained the main emporium for trans-desert trade for about two centuries. The Palmyra grew in importance as an autonomous state, and a main centre of trade, until its famous queen Zenobia was overthrown by the Romans in AD 273.

The Achaemenids, and the Sasanids after them, established a messenger service throughout the Empire, using relays of horse-riders. A series of posting stations were built, initially in the time of Cyrus (550–530 BC), each station staffed by a superintendent and grooms, and provided with relays of horses. The stations were placed one day's ride apart. The 2,600 km of the Royal Road could be covered by messengers in nine days, ten times as fast as an army. Nevertheless, these routes were not roads as we understand the term. They were simply tracks, no doubt chosen to follow the easiest routes,

Figure 5.1: Persian Road System

with masonry bridges at the more important crossings, but otherwise dependent upon fords.[2]

Greece

The topography of Greece is not favourable to the easy construction of a network of roads. The many mountain ranges of 3,000–5,000 feet in height cut the country up into sharply divided river valleys which are orientated towards the sea. The long, indented coastline and the numerous islands must have weighted the scales in favour of maritime traffic. Nor was the fragmentation of Greek society into city states, often at war with each other, conducive to the development of a country-wide road system.

One distinctive type of road that was built in Greece was intended for religious purposes. These so-called 'sacred roads' usually connect a city with a sanctuary in its neighbourhood, when the sanctuary was sufficiently important for frequent pilgrimages and religious festivals. The route is often the same as the route supposed to have been taken by a god on earth; for example the ceremonies on the road between Parnassus and Athens represent the journey of Apollo. It was essential that the procession and the four-wheeled carts with religious apparatus, statues of the gods or votive offerings should be able to proceed without shocks on a track as level as possible, while speed was not required. The construction of these roads was very unusual:

the track was roughly levelled and then artificial ruts were made from masonry, carefully hewn, polished and levelled to form a perfectly smooth and easy track for the cart wheels. Later on, this type of road came into restricted use for normal traffic. They appear, for example on the road between Athens and Piraeus. Sometimes they included passing-places, or even double rows of ruts; examples of double rows occur on the road from Athens to Delphi and on that from Sparta to Elis. The ruts are usually 7–10 cm deep, reaching a maximum of 30 cm in order to obtain a perfectly level track, with a width of 20 to 22 cm and a gauge of about 144 cm. These dimensions are similar to those on Maltese roads with artificial wheel ruts, which were built in pre-classical times; it is probable that the Greek roads were derived from these. The construction of these roads did not demand much engineering skill, since they usually followed a winding route along the natural contours.[3]

In most Greek cities, even at the time of their greatest prosperity, the streets were narrow, dirty and unpaved. Neither were they properly drained and this, together with the practice of throwing household refuse on to the streets, must have made them as insanitary and unpleasant as the streets of medieval European towns. Conditions improved somewhat in Hellenistic times, but it may well be that this was not due to the Greeks being converted to higher standards, but to the influence of eastern ideas. The earliest example of a paved street with proper foundations is Cyrene, in Libya, where a rammed limestone foundation course bore a stone slab pavement (about 635 BC). While the smaller streets in towns and cities were always narrow and dirty, the paving of main thoroughfares was a feature of Hellenistic cities. In Pergamon the main streets were paved with stone slabs, and Herod paved the streets of Antioch with marble. A special feature of Hellenistic towns, later adopted by the Romans, was the colonnaded street, the sidewalks of which ran under the arcades of the adjacent houses. Such streets were found in Alexandria, Seleucia, Ephesus and Gerasia among others. The main street of Palmyra was a colonnaded street 11.5 metres wide, while the colonnades were 5.5 metres deep; this thoroughfare was over 1,500 metres long and paved over the whole distance. An interesting feature of the construction is that the distances between the columns is gradually reduced, to give a false perspective and make the colonnade seem longer than it actually is.

Roman Roads

History

The origins of the Romans' skill in roadbuilding cannot be established with any certainty. It had been suggested that the Romans learnt something about roads from the Carthaginians, but this possibility has now been largely discounted. The Carthaginians were essentially a maritime people, who did not develop any system of roads to communicate with their hinterland — it seems likely that trade with the Berber tribes of the interior deserts was effected by the Berbers bringing their goods by camel train to markets in Carthage. There is no evidence for any roads, except some paved streets in Carthage and some local trackways. The pre-Etruscan inhabitants of central Italy seem to have had some knowledge of paving. In a settlement at Canatello, for example, paved streets from 2 to 3.8 metres in width were discovered. Little too is known about roads during the period of Etruscan domination (800–350 BC). The Etruscans seem to have a fairly extensive system of tracks, and some of the streets in their towns were paved. The town of Marzabotto possessed four main streets 15 metres wide with sidewalks. Most of the streets were paved and well drained, with stepping-stones to protect pedestrians against running water on the street surfaces. At Corneto the Romans are known to have found and rebuilt Etruscan paved streets and sewers, while in several towns the Roman paved roads followed the same course as the former Etruscan streets. Thus the Etruscan paved streets seem to have been much better constructed than their roads. As we have seen, paving was fairly common in Hellenistic cities, cities that were later incorporated into the Roman Empire.

It is a reasonable supposition that the Romans received their first ideas about the construction of roads from their Etruscan and Hellenistic predecessors, and that they developed their skills in the laying of paved roads by the accumulation of many years of experience.[4] The planning and construction of the road networks, however, must be attributed to a special talent for organisation and administration that was inherent in Roman society. Nothing like it had been attempted before, nor would be again before the eighteenth century.

All over the Roman Empire the roads were laid out as a carefully planned system linking the centres of occupation both military and civil, to every neighbouring centre, so as to ensure the most rapid communication possible. The roads were thus more nearly analogous to a railway system, and their layout was planned by well-trained engineers in much the same way, after a skilful survey of the ground problems to determine the most practicable route.[5]

It has sometimes been suggested that the purpose of the roads was primarily military, but this is not the case, except in some frontier areas such as the Danube and the Euphrates. In general, the permanent roads were not built until a newly-conquered region had been fully pacified, and they were intended to facilitate the needs of administrators, tax-collectors, and merchants, although their use by the military was always important and could at times become paramount.

The first of the great Roman highways intended to secure communications with newly acquired territory was the Appian Way or *Via Appia*, begun in 312 BC by the censor Appius Claudius. Its first section ran from Rome to Capua, and it was later extended south to Brindisi. Its surface was originally gravelled, paving being added in 295 BC. The equally famous *Via Flaminia*, which linked Rome with the Po valley, was completed during the censorship of Gaius Flaminius in 220 BC. Thereafter road construction continued almost without interruption in Italy and Sicily; it was extended into Dalmatia in 145 BC, into Asia Minor in 130 BC and into southern Gaul ten years later. Under the Roman Emperors of the first and second centuries AD the great network of roads took full shape, stretching from Hadrian's Wall in north Britain to the edge of the Sahara, and from Morocco to the Euphrates. After about AD 200 the pace of road building fell rapidly. By the fourth century the Roman state was finding it difficult even to maintain existing roads.[6]

Construction

Straight alignments were a characteristic of Roman roads, but are not invariable: if a road had to follow a ridge or a river it takes the most convenient and winding course. The real purpose of the straight alignments was for convenience in setting out. Sighting marks could be quickly aligned from one high point to another, with intermediate marks adjusted between, probably by the use of movable beacons shifted alternately from left to right until all were brought into line; it is noteworthy that Roman roads nearly always make important turns upon high ground at points from which the sighting could conveniently be done. Slight turns occur on intermediate hill tops where the road follows a straight alignment for a long distance, for example, Watling Street between Canterbury and Rochester. In general, however, the alignments were laid out with rigid accuracy for very long distances.[7] In Britain the Roman roads are recorded on Ordnance Survey maps, both those in their original state and those that have been overlaid by modern roads, and it is very easy, particularly in southern England, to find stretches of Roman roads which show the characteristic straight alignment.

The paving was always constructed with local materials, although suitable stones or gravel might be transported for a few miles. It is not therefore possible to select one type of construction as typical. On the general method of building, however, we have a description by Statius, poet laureate to the Emperor Diocletian (AD 81–96):

Now the first stage of the work was to dig ditches and to run a trench in the soil between them. Then this empty ditch was filled up with other materials viz. foundation courses and a watertight layer or binder and a foundation (literally a 'lap') was prepared to carry the pavement (*summum dorsum*). For the surface should not vibrate, otherwise the base is unreliable or the bed in which the stones are to be rammed is too loose.

At last the pavement should be fastened by pointed blocks and be held at regular intervals by wedges. Many hands work outside round the road itself. Here trees are cut down and the slopes of the hills are bared; there the pickaxe levels the rock or creates a log from a tree; there clamps are driven into the rocks and walls are woven from slaked lime and grey tufa. Hand driven pumps drain the pools formed by the underground water and brooks are turned from their course.[8]

Apart from the mention of wedges, whose purpose is obscure, this is a clear and lively discription of a roadmaking gang at work. The general principles of construction were as follows: once the route had been marked out the surface was cleared of vegetation and loose material and levelled. Drainage ditches were dug at either side, and the excavated material from these, if suitable, could be used in the foundations. On the subsoil the base course (about 20–30 cm thick) consisted of sand covered by a layer of mortar, or two or more rows of hand-laid slabs; the next course was a layer of cobbles or crushed stone grouted with lime mortar (30–50 cm); the third layer was a rich concrete (30–35 cm); the top layer was usually built of paving slabs,

Figure 5.2: Cross-sections of Roman Roads

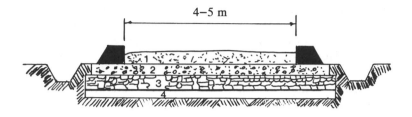

Normal Highway

1. 30–40 cm gravel concrete
2. 30 cm concrete with crushed stone
3. 30–50 cm slabs and blocks in cement mortar
4. 20–30 cm mortar layer on top of sand course

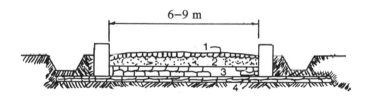

Main Highway

1. 20–25 cm cobbles or slabs in mortar
2. 30 cm concrete with crushed stone
3. 30–50 cm stone blocks in mortar
4. 20–30 cm flat stones

30–100 cm square laid on a bed of mortar (20–30 cm).[9] As indicated above, these materials could vary from region to region and also according to the classification of the road. Where hydraulic cement was available, as in Italy, this was used as the basis for concrete and mortar. A particularly effective form of surfacing was the hard slag from iron workings, which in some cases increased its effectiveness by rusting together into a very hard concrete-like mass.[10] It was usual, but not invariable, to add a top surface of fine gravel or chipped flint.

These then were the basic principles of construction. Although there were many variants, the two cross-sections shown in Figure 5.2 are representative of the techniques. For the highways the thickness was 100–140 cm, which is very great in comparison to modern thicknesses of 25–45 cm. Widths varied from 30 feet of roadway for important highways, down to 10–12 feet where terraces had to be cut into hillsides; a very common width was 15–18 feet. It has been estimated that the total length of the Roman roads in the days of the Empire was 50,000 miles,[11] so that it can be seen what a vast undertaking the creation of this system was. The material had to be quarried or excavated, transported to the site, prepared and laid — all this after the route had been surveyed, cleared and levelled. If we take all the Romans' public works together — roads, bridges, aqueducts, docks, harbours and public buildings — we must conclude that civil engineering absorbed a major part of the resources of the State.

Land Communications in Islam

The dromedary, or one-humped camel, was first domesticated in the second millenium BC in central Arabia; the first record we have of camel-mounted nomads occurs in the Old Testament in a reference to an incursion of the Midianites into Palestine in the first half of the eleventh century BC (Judges 6–8). There is ample evidence for the spread of dromedary culture throughout Arabia, Egypt and North Africa, Greater Syria, Iraq, and parts of Iran and Asia Minor by classical times.[12] The Bactrian, or two-humped camel had been known as a domestic animal in northern Iran since the third millenium BC, and its use spread out from that centre to the rest of Iran, Central Asia, China, northern India and northern Mesopotamia, reaching all of those regions by the first millenium BC.[13] In the areas just listed the two types overlap (in fact interbreeding is possible), whereas in the Arabic-speaking countries only the dromedary is used. It is better suited to dry, hot climates and it can be used as a mount for long-distance travel.

The donkey may have been domesticated as early as 4750 BC, and was certainly in widespread use as a pack animal in Islamic times. The horse was domesticated in the second millenium BC, but an incisive development in its use was the growth of 'equestrian nomadism' — to use Wissmann's expression. Perhaps about 900 or 800 BC, the Scythians and their neighbours, the Sakians, gave up steppe-farming life, probably in the region between the rivers Volga and Irtysh, and specialised in the breeding of herd animals, especially horses.[14] Thereafter, the breeding of horses on a large scale became an important element in the cultures of Central Asia and Iran. The introduction of the stirrup, probably in the seventh century AD, added greatly to the effectiveness of the horse as a mount, particularly in warfare.[15] Mules probably originated, also, in the second millenium BC and they were much used in the Islamic world. 'The postal service used these animals, and eminent men and women did not disdain to ride on them, in spite of their stubbornness and obstinacy, because their even gait and surefootedness made them valued mounts.'[16]

There was very little wheeled traffic in Islam; even in Spain the streets of the towns had to be readapted for vehicles after the Reconquest.[17] Animal transport was much more efficient, and in the desert areas of the Middle East and North Africa, the dromedary was the only possible means of long-distance travel until the introduction of motor transport. But the dromedary is more than just a means of transport. Without the specialised camel-culture of the Bedouin, life in the deserts and steppes would have been impossible, and would have been confined to isolated communities in oases.

Water-borne traffic was, of course, important in Islam. Merchant ships sailed the Mediterranean, and plied between the ports of the Persian Gulf and India. The rivers, such as the Nile, the Tigris, the Euphrates and the Guadalquivir were used extensively, as were the larger canals in Iraq and Sughd, for example. Even so, there were vast regions in which the only form of transport was four-legged. Islam in the Middle Ages was the most mobile society of any before modern times. The stimuli for this mobility were religious, commercial and political; the means were provided by the animal-breeding industries. The importance of the Pilgrimage to Mecca — the *hajj*— not only in fostering travel, but in the social structure of Islam, can hardly be overstated. The performance of the hajj is obligatory on every Muslim man or woman once in his or her life provided they have the means to do so. To quote Bernard Lewis:

The pilgrimage was not the only factor making for cultural unity and social mobility in the Islamic world — but it was certainly an important one, perhaps the most important ... The effect of the pilgrimage on communications and commerce, on ideas and institutions, has not been adequately explored; it may never be, since much of it will, in the nature of things, have gone unrecorded. There can, however, be no doubt that this institution — the most important agency of voluntary, personal mobility before the age of the great European discoveries — must have had profound effects on all the communities from which the pilgrims came, through which they travelled, and to which they returned.[18]

A second stimulus to travel came from commerce, which flourished in medieval Islam, especially from the ninth century to the twelfth. The geographers of the tenth century described this mercantile activity when it had reached its peak, in considerable detail. The range of goods was very large. It included foodstuffs of all kinds, many types of textiles, slaves, spices and dyestuffs, livestock, timber and manufactured metal goods. One passage from al-Istakhrī will have to suffice us as an example of this inland trade:

Isfahan was the trading centre in Jibal for Fars, Jibal, Khurasan and Khuzistan. It was the greatest exporting and importing centre in Jibal. It exported cotton and silk cloths to Iraq, Fars, Khurasan and elsewhere. Fruits from Nihawand went to Iraq, because of their excellence.[19]

There was also a lively trade with non-Muslim countries. The city of Mosul acted as a trading centre between the Byzantine Empire and Islamic countries,[20] and the port of Bab al-Abwab on the Caspian served a similar purpose for the exchange of goods with the people of the steppes.[21]

The postal service (*barīd*) was instituted by the Umayyad Caliphs in the seventh century, on the model of the Sasānid service, and was continued under the Abbasids. The organisation of the post is sufficiently well known for the Abbasid period from the works of the writers Ibn Khurradādhbih and Qudāma, composed for the secretaries of state in the ninth and tenth centuries.[22] There were no fewer than 930 stages in the Empire, situated, theoretically, at distances of 12 km apart in Iran and 24 km in the western provinces. The messengers used mainly mules in Iran, camels in the west, but horses were occasionally used. The postmasters had other responsibilities, besides the transmission of official letters. They had to provide central government with intelligence on the fiscal, economic and political situation in their provinces, and on the attitudes of important officials such as the commissioners for land taxation and the judges. From the tenth century onwards the service seems to have become increasingly disorganised until it was finally suppressed by the Seljuks in 1063.

In Mamlūk times, in Egypt and Syria, the postal service recovered its former importance for a while after it was reintroduced by the Sultan Baybars I (d. 1277). There was also a properly organised pigeon post, and a system of beacons stretching from Mesopotamia to Cairo, to give warning when the Mongols approached the frontiers. The invasion of Timur in 1400 destroyed this organisation.[23] A remarkable use of the postal service, said to have been introduced in Iraq in the eighth century, was the transport of ice for use in cooling drinks. In Mamlūk times five camel loads were sent at regular intervals from Syria to Egypt, at the expense of the Egyptian treasury.[24]

It was characteristic of Muslim towns that the markets and the mosques were in their centres, and it was only in this area that the streets were wide enough to allow public, commercial and social activities. The only other main streets were those connecting the central area to the main gates. These were strictly non-residential, and were deserted at night, when they were patrolled by guards. The residential sections of the town consisted of narrow lanes, one to three metres wide, many of them culs-de-sac. These could be locked by a gate at night. Drainage for rainwater and sewage was provided by channels in the middle of the streets.[25] Many of the side streets were probably left unpaved, but in Samarkand, for example, most of the streets, lanes and squares were paved with stone.[26]

Land Communication in Medieval Europe

It is generally accepted that a period of stagnation in Europe followed the end of the Roman Empire. In the tenth or eleventh century, after the pressures from the Vikings, the Muslims and the Magyars had relaxed, there began a period of economic growth in Europe that was to continue throughout the Middle Ages. The turning point in the European economy came earlier in Italy than in France, and earlier in much of France and the Rhineland than in the rest of Germany. The last areas to be touched by the expansive economy of the high Middle Ages were Scandinavia, east-central Europe and the interior of the Balkans.[27] Pilgrimages — to Rome, Compostella and the Holy Land

— were an important feature of medieval European life, although less important than in Islam.

The rivers of north-west Europe have a fairly even rate of flow throughout the year and are usually free from ice. Rivers such as the Po, the Seine and the Rhine became very important for transport in the early Middle Ages, but declined in the later part of the period, due mainly to the burden of tolls and other impositions levied by the riparian cities. One such imposition was a result of the staple rights of many European towns. They consisted in the privilege of forcing all passing ships to stop, unload, and offer their goods for sale, and to reload and proceed on their way only if there was no demand for them. It is very difficult for river traffic to evade such exactions, and the result was that more goods were transported on the roads. For personal transport people rode horses or mules — or simply walked. Goods were carried by human portage, by pack-animals, and in two-wheeled and four-wheeled carts.[28]

The Roman road system gradually fell into disrepair during the Middle Ages. The paved surfaces were expensive to maintain, but in any case, with the rise of new towns and changes to the trading routes, a new pattern of traffic emerged, the needs of which could only be met by developing new routes. More often than not, existing footpaths and bridleways were integrated into a multiplicity of tracks. There was no proper system for ensuring the upkeep of roads. In theory the maintenance of roads was an obligation on landowners, but this obligation was often disregarded. Hence compulsory labour was used, but this was not very satisfactory, since only in the towns was there sufficient control. Tolls were imposed with indifferent effect in some areas, but private donations were probably of greater import-ance. Municipalities, religious orders, private individuals and guilds were all active in promoting the construction and repair of roads, which was regarded as a pious duty, but by the thirteenth century the crown often took the initiative. The *strata publica* was part of the legislation in thirteenth-century France, and in English common law 'the King's highway' meant not only military roads but all highways leading to towns and markets.

The Roman road was an ideal road for pedestrian traffic, but it was not well suited to the transport needs of the Middle Ages. Gradually highways of cobbles or broken stone on loose foundations of sand, which could expand or contract with heat and cold and which was easily repaired, became the dominant type. The medieval state built *chemins ferrés* with blocks cemented with mortar, laid on sand foundations. There were also roads of gravel or broken stone on a basis of sand or earth.[79]

Notes

1. Herman Schreiber, *The History of Roads*, tr. Stewart Thomson (Barrie and Rockcliff, London, 1961), pp. 10-11.
2. R.J. Forbes, *Notes on the History of Ancient Roads and their Construction* (2nd edn, A.M. Hakkert, Amsterdam, 1964), pp. 80-3.
3. Ibid., pp. 96-113.
4. Ibid., pp. 115-21.
5. Ivan D. Margary, *Roman Roads in Britain* (John Baker, London, 1967), p.17.
6. R.J. Forbes, 'Evolution of Roman Roads' in Charles Singer, E.J. Holmyard, A.R. Hall and I. Williams (eds.), *A History of Technology* (5 vols., Oxford University Press, 1956), vol. 2, pp. 500-8.
7. Margary, *Roman Roads*, pp. 18-19.
8. Forbes, *History of Ancient Roads*, p. 139.
9. Ibid., p. 138.
10. Margary, *Roman Roads*, pp. 19-21.
11. Hans Straub, *A History of Civil Engineering*, tr. E. Rockwell (Leonard Hill, London, 1952), p. 5.
12. H. von Wissman and F. Kussmaul, 'Badw' in EI, vol. 1, pp. 881-9.
13. R. Walz, 'Beiträge zur ältesten Geschichte der altweltlichen Cameliden' in *Actes du 4é Congrès international des Sciences Anthropologiques et Ethnologiques* (Vienna 1952, publ. 1956), pp. 196-7.
14. H. von Wissman and F. Kussmaul, 'Badw' in EI, vol. 1, pp. 878-80.
15. Lynn White Jr, *Medieval Technology and Social Change* (Oxford University Press Paperback, 1964), pp. 24-5.
16. Charles Pellat, 'Baghl' in EI, vol. 1, p. 909.
17. Thomas F. Glick, *Islamic and Christian Spain in the Early Middle Ages* (Princeton University Press, Princeton, 1979), p. 119.
18. B. Lewis, 'Hadjdj' in EI, vol. 3, pp. 37-8.
19. Al-Istakhrī, *Kitab al-masālik wa'l-mumūlik*, ed. M.G. 'Abd al 'Al al-Hīnī (Cairo, 1961), p. 118.
20. Ibn Hawqal, *Kitāb surat al-ard*, ed. J.H. Kramers, 2nd edn, vol. 2 (BGA, Brill, Leiden, 1938), p. 215.
21. Ibid., pp. 382-3.
22. Extracts from *Kitāb al masālik* of Ibn Khurradādhbih and *Kitāb al-kharāj*, of Qudāma are in vol. 6 of BGA, ed. M.J. de Goeje (Brill, Leiden, 1889). It is worthy of comment that Ibn Khurradādhbih had himself been Director of Posts and Intelligence for a while.
23. D. Sourdel, 'Barīd' in EI, vol. 1, pp. 1045-6; for Mamlūk postal service see al-Qalqashandī, *Subh al-a'sha*, ed. M.A.R. Ibrāhīm (Dār al-Katub al-Khadiwiyya, Cairo 1913-20, 14 vols.), vol. 14, pp. 371-404.
24. Al-Qalqashandī, *Subh al-a'sha*, vol. 14, pp. 395-7.
25. Glick, *Islamic and Christian Spain*, pp. 114-5.
26. Al-Istakhrī, *Kitāb al-masālik*, p. 178.
27. N.J.G. Pounds, *An Economic History of Medieval Europe* (Longman, London, 1974), pp. 90-2.
28. Ibid., pp. 386-93.
29. R.J. Forbes, 'Roads and Traffic in the Middle Ages' in *A History of Technology*, ed. Singer *et al.*, vol. 2. pp. 524-7.

Plate 1 Cornalvo Dam

Plate 2 Muslim Weir at Valencia

Plate 3 Murcia Dam

Plate 4 Almonacid de Cuba Dam

Plate 5 Almansa Dam

Plate 6 Aqueduct Bridge – Aqua Claudia

Plate 7 Masonry of Aqua Claudia Bridge

Plate 8 Aqueduct Bridge — Mérida

6　Building Construction

Analysis of Structures

In order to discuss briefly the evolution of structures over a period of 2,000 years, we have no choice but to eliminate all the differences of style that are purely architectural, and to reduce buildings to their basic structural constituents. We also have to use modern technology when we make a statical analysis; this is acceptable so long as we bear in mind that the early builders did not think in this way, but solved their problems by intuition and experience.

Structurally, there are only four possible components in large masonry buildings: walls, columns, beams and arches. It is the way in which these components are combined, together with the materials of construction and the type of decoration, that distinguishes one architectural syle from another. The Greeks made very little use of the arch, relying almost entirely upon post-and-beam construction. Their temples consist of rows of columns which originally supported roofs made of massive timbers. The Romans, on the other hand, made widespread use of semicircular arches, domes and vaults, using beams only for lintels over doors and other openings. Their style of building came to be influenced by the use they made of good-quality concrete. Finally, the general application of the pointed arch in Europe from the twelfth century onwards, gave a new spaciousness to large buildings. It is customary to divide the history of architecture into well-defined periods: Greek, Hellenistic, Roman, Romanesque (Byzantine and Norman), Muslim and Gothic. Not all of these distinctions are meaningful from a structural point of view. Only the Greek period, with its absence of arches, and the Gothic, with its special use of the pointed arch, can be regarded as structurally somewhat different from the others. From about 200 BC to AD 1100 the main structural components of all large buildings were the semi-circular arch (the barrel vault is a continuous arch and the dome is an arch in the round), columns and walls.

The semicircular arch, which was known in Iran and Egypt long before Roman times, thus became, together with its near relations the vault and the dome, an important feature of most major buildings in Europe and western Asia. Its use poses certain problems, one of which is common to all arches, namely that at its junction with a side wall or column it exerts both a vertical and a horizontal thrust. The first causes no difficulties, because masonry is well able to take compressive stresses, but the horizontal thrust can lead to over-turning or produce tensile stresses in the supporting members and, as we have seen in the case of bridges, masonry cannot withstand even moderate tensile stresses. These difficulties may be overcome in various ways. The structure above the arches may be lightened by introducing a second row of arches in the upper wall or by making domes out of wood instead of masonry. The load-bearing walls may be made massive, although this makes for a ponderous effect, or they may be supported by buttresses. Arched aisles on either side of the main part of the building were also used to reduce the horizontal thrust imposed on the outer walls.

There is another disadvantage, this time peculiar to semicircular arches, in that their height is rigidly determined by their span. This caused great difficulties to the builders of medieval cathedrals as their height steadily increased and their ground plans became more elaborate. Something better than the semicircular arch was needed to lighten the superstructure and to give more flexibility to the designs. The innovation that made Gothic architecture possible was the pointed, or ogival, arch. This type of arch is one of the few cases we know of where a technological transmission from one culture to another can be traced with some confidence. The ogival arch is first found in Buddhist India in the second century AD and had reached Syria, possibly by way of Sasānid Iran, by AD 561. A number of such arches appear in Syrian buildings of the eighth century and it was common in Egypt in the ninth century. By about 1000 it appeared in Amalfi, which at that time had close commercial links with Fatimid Egypt,[1] and by 1071 a porch with pointed arches and vaults graced the new church at Monte Cassino. In 1080 Abbot Hugh of Cluny visited Monte Cassino, and as a result of his visit the new church of Cluny incorporated pointed arches and vaults. In 1130 Abbot Suger of the French royal abbey of Saint-Denis visited Cluny and between 1135 and 1144 he and his engineers produced at Saint-Denis what is usually regarded as the first true Gothic church.[2]

The structural devices which enabled the Gothic builders to cover a maximum of space with a minimum of material, and in particular to bestow upon the naves of their cathedrals such awe-inspiring heights, have been summarised by Straub as follows:

1. Distinction between bearing pillars and non-bearing walls, serving merely as enclosure (mostly windows).
2. General application of the easily adaptable pointed arch.
3. Distinction between vault-supporting ribs and intermediate panels of lighter weight.
4. Perfection of buttresses and 'flying buttresses' to absorb the vault thrust.

    We are thus confronted with a kind of framework structure. But, as all the members were merely able to absorb and transfer thrusts, these structures were, of necessity, much more intricate and accomplished than present-day systems which can also make use of tension members and stiff-jointed frames, in addition to pillars and struts.[3]

The pointed arch is statically more efficient than the semicircular type, and delivers its thrust at a steeper angle. Even so, the thrust still has two components, a vertical and a horizontal. The first, as is the case with all arches, is absorbed by the load-bearing column at its springing, high above the nave. The system for dealing with the horizontal component demonstrates the highly accomplished, albeit empirical, structural skills of the medieval engineers. The horizontal thurst is transmitted to the next column through a buttress or a flying

Figure 6.1: Gothic Building

buttress, the column absorbs the diminished vertical thrust while the diminished horizontal thrust is counteracted by the buttress-like thickening of the column. In the best Gothic work, the lines of thrust almost coincide with the structural members, but later the function of the framework was blurred by irrelevant decoration (Figure 6.1).

The Gothic system is thus a very elegant solution for the problem of distributing compressive loads, but its merits should not be allowed to obscure the fact that magnificent buildings have been erected by other means. Although we have classified Greek buildings as 'post-and-beam' structures, the 'posts' are actually beautifully made columns, perfectly finished and proportioned, which have left an indelible impression upon many generations of observers. From Roman times onwards, the dome has been used to give a sense of spaciousness and grandeur to the interiors of buildings. The Pantheon at Rome, which still survives, was built in the first quarter of the second century AD. It is surmounted by a dome built of brick and concrete, with the enormous span of 142 feet. It rests on a circular concrete wall, strengthened by a system of massive brick relieving-arches. In Byzantine times, and later, domes were made lighter, presumably in order to ease the problems of construction and to enable the supporting structures to be somewhat less massive. The great church of Hagia Sophia in Constantinople was built in less than five years, being completed in 537 — an amazingly short time for such a magnificent building. It is a flattish curve constructed of brick ribs, each 2 feet $4\frac{1}{2}$ inches wide at the springing, with lighter panels between each pair of arches. (A similar system was to be used in the vaulted roofs of Gothic cathedrals, except that the building material was stone.) The dome of the church of San Vitale at Ravenna, built between the years 526 and 547, when the city was part of the Byzantine Empire, is constructed from earthenware pots laid in tiers. The dome of the Dome of the Rock in Jerusalem, completed about AD 700, is made up of two wooden shells connected to each other by a system of bracing members.

Mosques, although they are closely related, structurally, to the Romanesque style have, of course, a very distinctive appearance. This is partly conferred upon them by the use of unique features such as the minarets and the prayer-niches (*mihrāb*), and also decorations in geometrical patterns or the Arabic script. It is, however, the blend of the large sunlit open spaces of the great courtyards with the silence and serenity of the colonnaded interiors that gives mosques their unmistakable character.

Materials

Bricks

Sun-dried bricks have been in use since Sumerián times and have always been a popular type of building material in the Middle East and the Mediterranean. They are relatively cheap, and even in humble dwellings their use permits the construction of thick walls, providing good insulation against heat and cold. As with many other matters, our earliest real information comes from Vitruvius:

> Therefore, first I will speak about bricks, and from what kind of clay they ought to be brought. For they ought not to be made from sandy nor chalky soil nor gravelly soil: because when they are got from these formation, first they become heavy, then, when they are moistened by rain showers in the walls, they come apart and are dissolved. And the straw does not stick in them because of their roughness. But bricks are to be made of white clayey earth or of red earth, or even of rough gravel. For these kinds, because of their smoothness, are durable. They are not heavy in working, and are easily built up together.[4]

He goes on to recommend that bricks should only be made in the spring and autumn, because if they are made at the hottest time of the year, the outsides will be baked hard but under this crust the insides will be soft. The best practice is to allow them to dry for two years; indeed, Vitruvius says that the citizens of Utica (about 25 miles from Carthage) used no bricks for building walls unless the magistrate had approved them as being dry and made five years before. According to Wulff, however, the practice in present-day Iran is completely contrary to the instructions given by Vitruvius. The bricks are made between May and October and are used after one or two days of drying 'for the building of common houses, for outside and inside walls, even for vaults and domes'.[5] It is difficult to reconcile the two statements; Vitruvius is certainly right in saying that very rapid drying would leave the inside of the bricks soft, but a drying time of two years — let alone five — seems to be excessive. The average thickness of Roman bricks was only about $1\frac{1}{2}$ inches, and they must surely have dried out in a short time. My own observations in Syria confirm Wulff's description, which is perhaps applicable especially to regions where the annual rainfall is low.

To prepare the mixture for the bricks, the loam or clay is

thoroughly soaked, mixed with straw and chaff, and trodden with the bare feet. It is then carried in baskets to the moulders. To quote Wulff (omitting the Persian words):

Each one has a wooden mold, just an open frame. The molder first covers the ground with a thin layer of chaff, puts the molding frame flat on the ground, and throws a quantity of the mud–straw mix into the mold, beats it into the corners with his bare hands, and scrapes any surplus off with a small straight–edge. He lifts the frame with a swift movement, leaving the fresh brick on the ground, and places the frame next to the brick just made. Molding row after row in this way, he makes about 250 bricks in an hour.[6]

In present-day Iran bricks are usually $8 \times 8 \times 1\frac{1}{2}$ inches, in Babylon they measured $16 \times 16 \times 4$ inches, in Sasānid times they were 15 to 20 inches long and $3\frac{1}{2}$ to 5 inches thick, and they were $9 \times 9 \times 2$ inches in early Islamic times.[7] Vitruvius gives three sizes of bricks, two of which were square and used by the Greeks, and a third oblong and used by the Romans; he gives its size as $1\frac{1}{2}$ feet long by 1 foot wide, but the Romans also used square and triangular bricks. The normal thickness for all of these was about $1\frac{1}{2}$ inches, but the 'Etruscan' bricks were $16 \times 10\frac{1}{2}$ inches and as much as $5\frac{1}{2}$ inches thick.[8]

Burnt bricks were already made in the fourth millenium BC in Babylonia, and in Persia kilns have been unearthed going back into the first millenium BC.[9] They do not, however, seem to have been used in Rome before the time of Augustus, and they are not described by Vitruvius. The preparation of the clay is more thorough than in the case of sun-burnt bricks — the clay has to be slaked and sieved to clean it of impurities, and additives are sometimes included, for example grey sand to give the bricks a whitish tinge. After moulding they are left in the open, lying flat, for 24 hours, and then turned on edge and left to dry for another three days before being stacked in the kiln. In Persia, a kiln for 50,000 standard bricks measuring $8 \times 4 \times 2$ inches would require a firing time of 72 hours.[10] Burnt bricks were widely used in the Roman Empire, but except in northern Italy and Byzantium they went out of use in Europe after the fall of the western Empire. Either from these areas, or from Islamic Spain, the use of brick spread again to southern France in the eleventh century and reached other parts of Europe, including eastern Britain, by the thirteenth. The municipal brickyard at Hull in the fourteenth and fifteenth centuries turned out about 100,000 bricks a year.[11] An

Arabic treatise of the eleventh century tells us that at Seville bricks and tiles were made outside the city. The brickmakers were required to produce bricks and tiles of various types: 'molars' and 'napes' for lining wells, others for floors, others for lining ovens and tiles called *asimiyya* for the roofs of monumental water-clocks. The whole range had to be kept in stock so that they were available when required.[12]

Stone

It is expensive to carry stone for long distances, and the type of stone available in a region has therefore usually determined what methods of building construction will be used in that region. For example, buildings in the alluvial plain of Iraq have always been made of brick, because stone is not available locally. Building stone, chiefly limestone and marble, is abundant in Greece, and important buildings in the classical period were of one or other of these materials. The limestone was of the variety called *poros*, found in the west and north of the Peloponnese. This stone had a rough surface and was full of cavities, thus providing an excellent key for plaster. It was often plastered over with fine stucco, inside and outside a building, and then colour washed. The marble most often used in Greek buildings, Pentelic marble, was and is quarried at Mount Pentelicus, a few miles north of Athens. It consists almost entirely of calcium carbonate, showing a brilliant white fracture when broken. This material was employed for the Parthenon, the Erechtheum, the Propylaea, the temple of Zeus Olympios, and most of the other principal public buildings of Athens.[13]

The most popular building stone in Rome was travertine, which was chiefly quarried near Tivoli. It is a cream or black limestone from deposits in the valleys of the Tiber and Aniene. It varies greatly in texture, being sometimes fine, sometimes coarse with numerous cavities, which afforded excellent bonding for plaster. Tufa is a very different kind of stone, being of volcanic rock, resembling pumice. It was used in Rome, mostly before the time of Augustus. It was laid in large blocks, and had to be protected externally by rendering. Peperino and Alban stone were also used.[14] Vitruvius lists several places in central Italy where the quality of the stone was excellent. If it was necessary to obtain stone from quarries closer to Rome, he recommends that after the stone had been excavated, it should be left exposed for two years; any stone that had been damaged by weathering should be thrown into the foundations. Marble was not introduced into Roman building until the first century BC and was rare before Augustus. Unlike the Greeks, who used only white marble, the Romans also imported many coloured foreign varieties. Granite was imported from Egypt, Elba and Naxos. Great monolithic columns were made from red basalt, known as porphry, which was brought from Egypt.[15]

It would of course be impossible to list all the types of stone used in the masonry structures of Europe and the Middle East, particularly when we consider the wide variety of stone used in England alone, for example, from Roman times onward.[16] But the availability of good stone does not necessarily mean that the stone will be used to construct good buildings in the area where it is found. Economic and political factors, or a shortage of skilled labour, may lead to the use of low-quality materials and techniques. Only in certain areas, one being the Cotswold region of England, has a tradition of sound building in local stone been maintained over long periods. In Syria also, while poorer methods of construction were in use in other parts of Islam, buildings made from dressed masonry in the beautiful local limestone continued to be erected.

Mortar and its Constituents

Nowadays we think of mortar as having two main purposes: the bonding of masonry and the plastering of walls. It served the same purposes in earlier times, but it was also considered as a separate constituent in the manufacture of concrete.

The Romans did not mix batches of concrete consisting of cement, sand, aggregate and water before placing it in the shuttering, as is the modern practice; they pre-mixed only the cement and sand, with water, and then poured this mortar on to fillings of rubble. This practice would not be approved by modern civil engineers, but the fact is that, for many duties, the compressive strength of concrete is more than adequate for carrying the loads, as is testified by the many surviving Roman structures made of concrete.

The specifications given by Vitruvius for sand are in accordance with modern practice. Sand to be used for concrete should preferably be obtained from sand-pits, it should make a noise when rubbed in the hand and it should be free from earthy impurities. Sand sifted from rivers is less suitable, since it is not strong enough for use in vaulting. Sea-sand should not be used for plastering, because the salts in it will effloresce from the walls. Fresh pit sand, while it is ideal for concreting, is unsatisfactory for plastering, since it will not provide a smooth crack-free finish. River sand, which can be worked to a smooth finish by trowelling, should be used for this purpose.[17]

Lime was the most important ingredient of mortar in classical and medieval times. It is made by 'burning' chalk or limestone in kilns and then slaking it with water. It was mixed with sand in the proportion of one to three for pit sand, one to two for river sand or sea sand. Mortar suitable for jointing and plastering was also made from gypsum; al-Idrīsī tells us that the mountains near Almeria were composed of gypsum, which was transported to the city for use in buildings.[18] Pozzolana was a most important mineral found in thick strata in the Alban Hills and near Naples. This was a volcanic earth that was mixed with lime to form a mortar that set strongly under water and was also fire-resistant. Its use made possible the construction of the impressive vaults and domes over the public baths, and it was a valuable material for the piers of bridges, and in harbour works.[19]

Timber

The countries of the Mediterranean have suffered from progressive deforestation over the centuries; timber was much more abundant in classical times, and even in the Middle Ages, than it is today. In northern and western Europe, timber was readily available until fairly recent times. It was certainly a very valuable structural material, both for important buildings and humble dwellings, but in comparison with masonry so little of it has survived, that we must rely mainly upon literary sources for our information. Vitruvius discusses extensively the different types of timber and their uses in building. Briggs summarises his discussion as follows:

> He recommends felling during winter, when the leaves have withered and, he thinks, the roots having drawn sap from the soil recover their solidity. He gives advice about the conversion of the tree into planks, 'so that the sap may dry out by dripping'.
>
> As regards the types of timber, Vitruvius states fir is naturally rigid, and does not deflect much when used for floorings; however, it burns easily. The lower part of the trunk is generally quartered, the sapwood being rejected and used for inside work. Oak (*Quercus robur*) resists moisture well, even when buried in foundations. Winter oak (*Q. aesculus*), Turkey oak (*Q. cerris*), and beech are liable to decay. Poplar, willow, and lime are soft, white, porous, and suitable for carving. Alder had the remarkable quality of being 'imperishable underground', and will 'uphold immense weights of walling and preserve them without decaying

... All buildings at Ravenna, both public and private, have piles of alder under their foundations .' Cypress and pine, containing an excess of moisture, tend to warp but last long without decay, and the oil in cypress resists dry-rot and worm. Larch contains a bitter sap which also resists dry-rot and worm, but it is a poor fuel and so heavy that it will not float.[20]

Timber was widely used throughout our period for domestic building, but it was less important than the various forms of masonry as a structural material. It fulfilled a structural purpose, apart from its use in piling, only in the roofs of Greek temples, a few Islamic domes and in the trussed roofs over the vaults of some medieval cathedrals. It was indispensable for temporary structures such as scaffolding and the wooden centring for the construction of masonry arches. It was the normal material for doors, and was used to great effect in decoration, as can be seen, for example, in the carved screens in European cathedrals. In mosques the pulpits (*minbār*) are often fine examples of the art of the woodcarver.

Cobwork

The technique of cobwork, or rammed earth, is very old and may represent a stage in the evolution of true concrete, but its use is especially associated with Muslim building. Earth with which chalk and crushed baked earth or broken stones are often mixed is rammed between two boards kept in parellel by beams. The wall is plastered over, often in such a way as to simulate the joints of heavy ashlar masonry beneath. In the Muslim West cobwork became general in the eleventh and twelfth centuries, particularly for military building. In North Africa it seems to have been an import from Andalusia, where it had long been known.[21]

Metals

Although they were not generally used as structural members, metals served a variety of purposes in early buildings. The Greeks seem to have used iron with confidence in a variety of applications — in the hooks and tongs of lifting devices and for many kinds of clamps and dowels. They were also exceptional in their use of iron as a structural material. In the Parthenon, broad flat wrought iron beams were used as cantilevers to support the heaviest statues of the pediment, their ends being built into the masonry of its recessed face. Special provision was made so deflection would not cause them to bear directly on the marble cornice. In one temple of about 470 BC at

Agrigento, iron beams of 5 × 12 inches in cross-section and 15 feet long were let into the under-surfaces of the architraves and rested on the column-capitals. In another temple at the same place, iron cantilevers carry the topmost member of the cornice. In the Propylaea at Athens concealed iron beams 6 feet long transmitted loads from the heavy marble ceiling-beams on to the Ionic columns. Here, too, due allowance was made for deflection.[22] As far as is known, the successors of the Greeks did not use iron in this way, but it continued to be used for many fixings and fittings. The decorative capabilities of wrought iron were fully realised in the intricate ironwork screens which can still be seen in many cathedrals and in Muslim palaces and mosques.

The use of lead as a jointing material was a common practice in Greek, Roman, Romanesque and Islamic buildings. The commonest method was to pour molten lead into dovetail joints between blocks of dressed stone, but the use of lead as mortar in dressed stone masonry — a somewhat wasteful practice — was known in Islam.[23] Sheet lead as a roofing material came into early use in England and France. York Minster is said to have been roofed with lead about 669, as was the choir of Canterbury Cathedral (1093-1130) and, somewhat later, the aisles. At Chartres Cathedral in the thirteenth century the sheets were about one sixth of an inch thick, and not more than 2 feet wide and 3–4 feet long. Several lead-covered spires of the Middle Ages still survive in England, as at Godalming and Chesterfield, both of which have the lead rolls fixed diagonally. Medieval lead was always cast, not milled; the casting was done on a bed of smooth sand, specially prepared for each sheet.[24]

Bronze was used for similar purposes to iron, for example, in clamps and dowels for masonry, in locks and in the fittings and hinges of doors. Sometimes bronze was used more ambitiously, as in the magnificent doors of the Pantheon in Rome, which are still in existence. Brass and copper were used mainly for decorative purposes (and very extensively for domestic utensils), but in the work of al-Jazarī (composed AD 1206) there is an illustrated description of a beautiful double-leaved door, made of brass and copper, for the palace at Amid in Upper Mesopotamia. Many of the decorative motifs were made by the lost-wax process, which was known in classical times, but the central part of each leaf was made from brass, using closed mould-boxes with green sand. No earlier description of this technique is known — it was first introduced to Europe towards the close of the fifteenth century. No trace of these doors remains, but they must have been an impressive sight — each leaf was about 15 feet high and 5 feet wide.[25]

Methods of Construction

We have a good deal of information, mainly from literary sources, about the methods used for transporting and lifting heavy loads. Otherwise, although we known the names of the architects of some early buildings,[26] we really know very little about how architects conceived and executed their work until we come to the builders of the Gothic period. For this period, recent researches have given us some insight into the methods of English architects.[27] It is reasonable to infer that other architects, at different times and in different places, used similar techniques for translating their ideas into reality.

Moving Heavy Loads

Classical authors are unanimous in attributing to Archimedes the introduction of the multiple pulley and other methods of combining the five basic machine elements to obtain maximum mechanical advantage. Vitruvius has a good deal to say about lifting methods,[28] as does Hero of Alexandria (*fl. c.* AD 65), whose Mechanics is known only from an Arabic translation of the ninth century.[29] In his scientific encyclopaedia, *The Keys of the Sciences*, written towards the close of the tenth century by Abu 'Abd Allah al-Khuwārizmī, there is a section on moving heavy weights that incorporates much of Hero's material.[30] None of these writers was describing anything radically new; no doubt they based their descriptions upon working practices and the writings of Archimedes, who himself probably developed his ideas from age-old Egyptian methods (he is known to have lived for some time in Egypt). Nevertheless, we must be grateful to these writers, since without the information contained in their works we should know very little about the handling of materials in early times.

Some of the loads were very large. The drums of the columns in the Parthenon are over 6 feet in diameter, those at Selinus in Sicily 10 feet $8\frac{1}{2}$ inches across. Three great blocks of stone in the terrace of the Roman temple of Jupiter at Baalbek in the Lebanon measure 63–65 feet by about 13 by about 10. It was usual for roadways to be prepared from the quarries to the building sites and for the loads to be hauled along these tracks in wagons by teams of oxen. Blocks of stone of square or rectangular cross-section were sometimes enclosed by wooden 'wheels' so that they could be rolled along the ground (Figure 6.2a). Vitruvius relates that the shafts of the great columns for the temple of Diana at Ephesus had to be transported along a flat but soft stretch of ground where wagons could not go. So the contractor, Chersiphron, put iron staples into the ends of each column, made a frame long enough to reach from end to end, and had the columns dragged by oxen, like enormous garden rollers.

The attachment of hooks of lifting devices to the loads presented a problem that was solved in various ways. For example, Hero describes a wedge-shaped piece of iron with a hole bored through it near its apex. This fitted into an undercut hole in the masonry block and was held in place by a block of wood (Figure 6.2b). In the column-drums of some Greek temples, rough bosses of marble, 8–10 inches square, have been left projecting for some 6–8 inches at four points on the circumference. These were probably lifting trunnions: a lifting strop with a loop at either end would have been placed over each protuberance, while the loops at the other ends were placed over the hook of the lifting appliance (Figure 6.2c).

The commonest type of crane, as described by Vitruvius, was the sheerlegs. Two stout spars were held apart at the top by a short cross-piece; at the bottom a windlass was installed on bearings fixed to the legs; the lifting tackle was attached to the cross-piece by a loop of rope — it was a block-and-tackle, the upper block having two

Figure 6.2: Lifting Devices

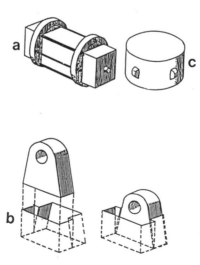

sheaves, the lower one sheave (a three-and-two was also used). The rope from the block and tackle was wound on to the drum of the windlass. The crane was held in position by guy ropes attached to the tops of the spars and to stakes driven into the ground (Figure 6.3).[31] The sheerlegs can only be moved in one plane and single derricks were used when it was necessary to swing the loads through an arc. Vitruvius says that for heavier loads a lifting device may be operated by tread-wheels, and indeed there is a relief in the Lateran Museum showing a derrick operated by an enormous tread-wheel worked by five men. Compound pulleys are used in the stays as well as in the lifting tackle.[32]

Tread-wheels were also used in the construction of large medieval churches. In England, three of these are still in position in the towers of Peterborough Cathedral, Tewkesbury Abbey and Salisbury Cathedral. The first two are twelfth century constructions, while the one at Salisbury, although it was probably installed for the original construction, which began in 1220, may have been modified later to

Figure 6.3: Sheerlegs

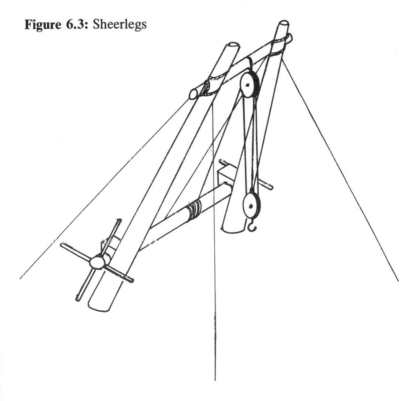

meet the requirements of the fourteenth-century spire. The main dimensions (diameter of wheel, diameter of barrel and length of barrel) are as follows: Peterborough — 9 feet 6 inches, 13 inches and 12 feet; Tewkesbury — 11 feet 11 inches, 16 inches and 6 feet 5 inches; Salisbury — 10 feet $10\frac{1}{2}$ inches, $10\frac{3}{4}$ inches and 5 feet 9 inches. In each case, iron caps carrying iron spigots are fitted to both ends of the barrel; these turn in iron journals sunk into timber posts.[33] By the fifteenth century, in Europe, cranes incorporated a swivelling mechanism.[34]

Execution of the Work

One of the most important pieces of documentation on large building and engineering projects nowadays is the 'organisation chart', which shows the interrelationships among the members of the project team — project managers, architects, consultants, specialist engineers, surveyors and so on. No such document was needed in the Middle Ages, because the master mason fulfilled all the roles that are today divided up among various disciplines. No single definition of the duties of a master mason will suffice because these could vary from one project to the next. His activity could be in four major areas: as architect of the building; as technical supervisor of construction; as administrative official; as building contractor. It is the first two of these functions which concern us here.

The education and training of the medieval master mason were quite different from those of the present-day architect. Although the masons were by no means all illiterate, they did not learn from books, mathematics and architectural drawings but rather through the direct teaching of full experienced masters. Shelby makes a cogent statement about the results of this method of learning:

> From this system of education there derived important conse-
> quences for the art of designing buildings. It meant that the
> architects — the master masons — worked within the framework of
> an organic tradition, that of the mason's craft. It is probably for this
> reason that mediaeval designers seldom indulged in radical
> innovations, nor did they seek artificial renaissances of the
> architectural standards of bygone ages. There were changes in
> design, but they were continuous and evolutionary, rather than
> radical or reactionary. Furthermore, since mediaeval architects of
> stone buildings were themselves masons, they knew their building
> material at first hand. This helps to account for the great skill of

these designers for both decorative and structural purposes. Their ability in combining form with function, particularly in the greater cathedrals and churches, has been the source of much wonder and praise by architectural historians. Yet it should occasion no surprise that the great mediaeval tradition of building in stone was developed in a period when the architects themselves were thoroughly imbued with an empirical knowledge of the materials and forms available to them for the expression of their architectural ideas. Finally, the training and background of the mediaeval designer meant that there was not that separation between the architectural idea and the execution of the idea that is characteristic of modern building, for the relationship between the master and his masons was altogether different from that between modern architects and construction workers. This becomes evident when we compare mediaeval with modern techniques of transmitting architectural ideas from the designer to the workers who execute the design.[35]

Nowadays the architect prepares a complete set of working drawings and specifications that are transmitted to the builder who is to execute the work. The design data will have included the results of a soil survey, analysed by the methods of Soil Mechanics, so that even the design of the foundations should not need to be modified after the site is excavated. In theory the architect should not need to visit the site, except to deal with contractual matters, but in practice some visits will be necessary to deal with unforeseen problems that always arise during the course of construction. But these visits should not have to be made frequently. In the Middle Ages things were quite different. Drawings were sometimes made, but these were undimensioned illustrations to assist the clients to visualise the appearance of the completed buildings — what we should call artists' impressions. These were not intended to be used for passing on the architect's design to the builders. For this function there was no alternative to the constant supervision of the architect himself. His first task was to set out the ground plan of the building, and supervise the clearance and preparation of the site. In the process of setting out the work ropes, cords, fir poles and lime were used to mark the boundaries of the building as well as to indicate foundation lines for interior walls and pillars. Modern methods of testing the soil were unknown, so the choice of footing could often only be made after the excavations were under way.[36]

The supervision of the masons in the task of cutting the stones for the intricate designs of doors, windows, arches and vaults was achieved by the use of wooden moulds. There was often a special tracing room set aside for the master mason to draw the design on boards, which were then cut to shape by the carpenters. Masons would then start work, using the moulds to cut the stones to the desired shape and size.[37] Presumably an adequate supply of cut stones was always kept in stock, so that the work of erection could proceed without delays.

Notes

1. A.O. Citarella, 'The Relations of Amalfi with the Arab World Before the Crusades', *Speculum*, 42 (1967), pp. 299-312.
2. Lynn White Jr, *Medieval Religion and Technology* (University of California Press, 1978), pp. 231-3.
3. Hans Straub, *A History of Civil Engineering*, Eng. tr. by E. Rockwell (Leonard Hill, London, 1952), pp. 38-9.
4. Vitruvius, *De Architectura*, ed. and trans. Frank Granger (2 vols., Loeb Classics, London 1931, this reprint 1970), vol. 1, bk 2, Ch. 3, pp. 89-95.
5. Hans E. Wulff, *The Traditional Crafts of Persia* (MIT Press, Cambridge, Mass., 1966, 2nd printing 1976), p. 110.
6. Ibid., pp. 109-10.
7. Ibid., p. 110.
8. Martin S. Briggs, 'Building-Construction' in Charles Singer, E.J. Holmyard, A.R. Hall, Trevor I. Williams (eds.), *A History of Technology* (5 vols., Oxford University Press, 1956, reprinted 1979), vol. 2, Ch. 12, p. 408.
9. Wulff, *Traditional Crafts*, p. 115.
10. Ibid., p. 116.
11. G.M.A. Richter, 'Ceramics: Medieval' in Singer *et al.* (eds.), *A History of Technology*, vol. 2, Ch. 8, p. 305.
12. E. Levi-Provencal, *Trois Traités Hispaniques de Hisba* (Cairo, 1955), p. 34. (These are the texts of three Arabic treatises; the reference is to the text of Ibn 'Abdūn of Seville, who lived during the second half of the eleventh century and the first half of the twelfth.)
13. Briggs, 'Building-Construction', pp. 398-9.
14. Vitruvius, vol. 1, bk 2, Ch. 7, pp. 105-11.
15. Briggs, 'Building-Construction', pp. 406-7.
16. C.N. Bromehead, 'Mining and Quarrying to the Seventeenth Century' in Singer *et al.* (eds.), *A History of Technology*, vol. 2, Ch. 1, p. 32.
17. Vitruvius, vol. 1, bk 2, Ch. 9, pp. 131-45.
18. Al-Idrīsī, *Description de l'Afrique et de l'Espagne*, Arabic text with Fr. trans. by R. Dozy and M.J. de Goeje (Brill, Leiden, 1866), p. 201 Ar./245 Fr.
19. Vitruvius, vol. 1, bk 2, Ch. 6, pp. 101-5.
20. Ibid., Ch. 9, pp. 131-45.
21. G Marçais, 'Binā'' in EI, vol. 1, pp. 1226-9.
22. Briggs, 'Building-Construction', pp. 400-403.
23. Al-Istakhrī, *Kitābal-masālik wa'l-mamālik*, ed. M.G. 'Abd al-'Āl al-Hīnī (Cairo, 1961), p. 36.

24. Briggs, 'Building-Construction', pp. 429, 445.

25. Al Jazari, *The Book of Knowledge of Ingenious Mechanical Devices*, trans. Donald R. Hill (Reidel, Dordrecht, 1974), pp. 191-5.

26. C. Diehl, 'Byzantine Art' in N.H. Baynes and H. St L.B. Moss (eds.), *Byzantium* (Oxford Paperbacks, Oxford, 1961), Ch. 4, p. 166. Diehl gives the names of the architects of Hagia Sophia — Anthemius of Tralles and Isidore of Miletus, both from Asia Minor.

27. See especially: J. Harvey, *English Mediaeval Architects* (London 1954); and two papers by L. R. Shelby, 'The Role of the Master Mason in Mediaeval English Building, *Speculum*, 39 (1964), no. 3, pp. 387-403, and 'The Geometrical Knowledge of Mediaeval Master Masons', *Speculum*, 47 (1972), pp. 395-421.

28. Vitruvius, vol. 2, bk 10, Ch. 2, pp. 279-93.

29. Carra de Vaux, 'Les Méchaniques ou l'Élevateur de Heron d'Alexandrie sur la Version de Qostā ibn Lūqā', *Journal Asiatique*, ninth series (1893), vol. 1, pp. 386-472, vol. 2, pp. 152-269, 420-514. For the translator see Donald R. Hill, Kustā b. Lūkā in EI, vol. 5, p. 529.

30. Al-Khuwārizmī, *Mafātīh al-Ūlum*, ed. G. van Vloten, (Leiden, 1895).

31. Very similar devices were used by army engineers in the Second World War.

32. A.G. Drachmann, 'A Note on Ancient Cranes' in Singer *et al.* (eds.), *A History of Technology*, vol. 2, pp. 658-62.

33. W.G.C. Backinsell, *Medieval Windlasses*, Historical Monograph 7 (South Wiltshire Industrial Archaeology Society, Salisbury, 1980).

34. Bertrand Gille, 'Machines' in Singer *et al.* (eds.), *A History of Technology*, vol. 2, pp. 629-58, p. 657.

35. L.R. Shelby, 'Role of the Master Mason', p. 389.

36. Ibid., p. 401.

37. Ibid., p. 393.

7 Surveying

The techniques of surveying are required for all the types of construction discussed in the preceding five chapters: for levelling canals, qanats and tunnels; for the lines and levels of dams and bridges; for the alignment and levelling of roads; for the foundations of buildings and setting out the lines of their walls and columns. Another application of surveying, particularly in Roman times, was the division of land into square plots, for example in the allocation of holdings in newly-founded settlements to army veterans. In all these applications, accurate measurements within the limits of the available instruments, together with a knowledge of plane geometry, sufficed to provide satisfactory solutions for the problems. Indeed, the degree of accuracy, from earliest times, was quite astonishing. The Great Pyramid in Egypt, built in the third millenium BC, has a base plan 756 feet square, which comes within 7 inches of forming a perfect square, and the sides are oriented to within less than one tenth of a degree.[1] Survey by triangulation, to determine heights and distances by the use of right-angled triangles, requires a knowledge of plane trigonometry. Geodetic surveying, for example to determine the co-ordinates of places with reference to the equator and a prime meridian, is the most demanding of all from a mathematical point of view, since it involves the use of astronomy and spherical trigonometry. In any type of surveying, however, the final results depend upon the accuracy of the observations.

Instruments and Techniques

The minimum requirements for a survey of any kind are a level of some kind, a method for sighting along straight lines and a measuring line. Some method of orienting one's position with regard to the cardinal points may also be necessary. For the more elaborate kinds of survey an instrument for measuring horizontal and vertical angles is needed. Most of our information about instruments, and indeed about surveying in general, comes from Roman and Islamic sources, but we can infer that the tradition for accurate surveying remained unbroken from ancient times through the Greek period. Existing Greek structures are evidence enough, and we also know that some Greek towns were laid out on a grid pattern. For example, a certain Hippodamus of Miletus redesigned the Piraeus on this pattern after the Persian wars, and rebuilt his own city of Miletus on a system of squares in 479 BC.[2] As in other fields: 'What the Romans did was to combine features from Egypt, Etruria, Greek towns and Greek countryside to make their own distinctive system.'[3]

Levelling

Vitruvius describes an instrument which he calls a *chorobates*; this is simply a large builder's level consisting of a straight piece of wood 20 feet long, supported on legs, the joints being strengthened by knee-braces. There were plumb lines at each end, and when these lines coincided with marks on the braces the level was horizontal (Figure 7.1c). If the wind prevented the bobs from coming to rest, a groove on top of the level was filled with water to indicate when the beam was horizontal.[4] This is a cumbersome device, which could hardly have been used for cross-country work, but then the main concern of Vitruvius was with building construction. We may conjecture that one or more chorobates were set up on a building site to act as bench-marks from which other levels on the site were derived.

In the treatise called *Kitāb al-Hāwī*, written in Iraq in the eleventh century by an unidentified author, there are instructions for the manufacture and use of three levelling instruments. The first of these is a wooden board about 70 cm long by 8 cm wide. In the centre of the board a line is drawn which meets both edges at a right angle. A plumb line is fixed to this line, near one of the edges. Two hooks are fixed to this edge. The second instrument consists of an equilateral triangle made of metal, with two hooks soldered to either end of its hypotenuse. A narrow hole is drilled through this side, to take the cord of a plumb line. The third instrument is called the 'reed-level'. This consists of a straight reed through which a narrow longitudinal hole is bored. In the centre, a radial hole is made into the bore.

To use any of these instruments, two wooden levelling staffs are needed, each at least as tall as a man; the staffs are divided into *qabda* (lit. fist — about 12 cm), and each qabda is subdivided into finger-widths (about 2 cm). To use the first, one takes a cord 7.0–7.5 metres

Figure 7.1: Surveying Instruments

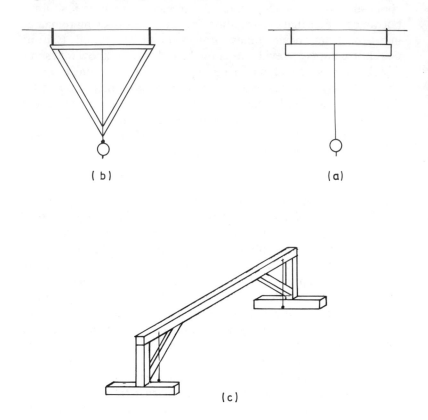

(b)

(a)

(c)

long and passes it through the hooks, bringing the board to the centre of the cord. One man takes one of the staffs and holding it vertically (no doubt with the aid of a plumb line) places the end of the cord against it. A second man takes the other end of the cord and drawing it taut, holds it against the second staff, which must also be vertical. The surveyor then checks the level, having one end of the cord raised or lowered until the plumb line coincides with the line on the board. The readings on both staffs are then noted down, and the difference recorded, either as a 'rise' or a 'fall'. At the end of the survey the algebraic sum of the rises and falls gives the difference in level between the start and finish points (Figure 7.1a). A similar method was used with the second instrument; in this case the triangle was suspended to the cord by its two hooks, and when horizontal its plumb line ran through the apex (Figure 7.1b). In the third method a man held each end of the tube against the staff, while a third man with a container of water and a rag of cotton or wool stood at the centre and squeezed water from the rag into the central hole. When the water issued from either end in equal quantities the tube was horizontal.[5]

Alignment and Triangulation

The *groma* was the standard instrument used by Roman surveyors for the process known as *centuriation,* the dividing up of land into squares and rectangles. The period following the civil war, when the Second Triumvirate had to provide land for thousands of discharged soldiers, was one of enormous expansion for surveying of this kind, and of growth of a body of surveyors known as the *agrimensores,* whose main duty was the carrying out of centuriation.[6] The fact that the groma could only be used for setting out straight lines and right angles was therefore no drawback, since this was exactly what the agrimensores were required to do. It consisted of a wooden staff over which, at the top, a metal bracket was fitted. An iron cross-piece with four arms fitted into the socket of the bracket, and at the end of each arm of the cross a plumb line was suspended. At the bottom of the staff was a pointed iron shoe to enable the groma to be firmly

Figure 7.2: Groma

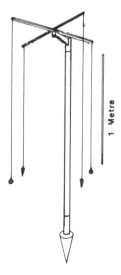

1 Metre

implanted in the ground. The purpose of the bracket was to move the centre of the cross off the centre-line of the staff, so that an unimpeded sight could be taken through either pair of plumb lines (Figure 7.2).[7]

The word *dioptra* in Greek means any instrument for taking a line of sight. In a treatise on surveying written by Hero of Alexandria about the middle of the first century AD, there is a description of a dioptra that is very similar to a modern theodolite apart from the absence of a telescope. The instrument is mounted on a column which is checked for verticality by a plumb line. At the top there is a bar provided with sights, and this can be turned in the vertical plane by a worm-and-pinion gear and in the horizontal plane by another such gear. The sighting bar could be replaced by a water-level. Although this is a very ingenious instrument with obvious relevance in the history of surveying, it was probably not very robust, and would soon have gone out of adjustment if subjected to the wear-and-tear of use on construction sites. For everyday use in measuring angles, a simple rotatable alidade provided with a scale of angles was probably the normal instrument.

The astrolabe, as an instrument for astronomical observation and computation, will be described in Chapter 10. Here we are concerned only with the back of the instrument, which consisted of an alidade turning about a central pivot, its ends moving over a graduated circle, each quadrant of which was divided into 90 degrees. In the lower half of the face was a rectangle, one side of which was divided radially into decimals, the other side into duodecimals. The alidade, which was usually of the shape shown in Figure 10.4d, was provided with sighting holes. (There were other markings on the back of the astrolabe, but they are not relevant to this discussion.)

A number of Arabic writers, including the great scientist al-Birūnī (d. c. 1050) and the Spanish Muslim Ibn al-Saffār (d. 1035), describe the solution of various triangulation problems using the astrolabe. The two matching squares into which the rectangle is divided are used for this purpose. Although the squares are divided into ten and twelve equal parts respectively, the choice of number is purely a matter of convenience. With the astrolabe freely suspended the alidade is adjusted so that the distant object is viewed through the sights simultaneously. When this happens the real right-angled triangle formed by the distance of the object and its height is reproduced on a small scale within the confines of one square on the astrolabe, by an exactly similar right-angled triangle. The real and similar triangles

Figure 7.3: Back of Astrolabe

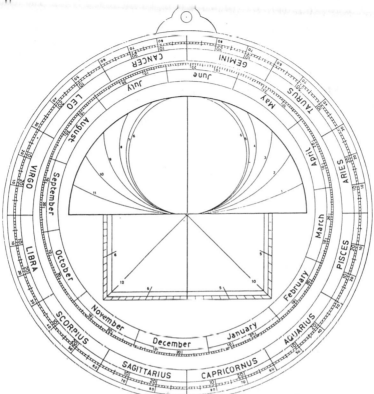

have a common line for hypotenuse. The ratio of the lengths of the sides of the triangle on the astrolabe is the same as the ratio of the height and distance of the object, so that if either of these is known the other may readily be calculated. If neither is known, a sight is taken on the top of the object, the angle is then decreased by moving the alidade through one decimal or duodecimal division, then moving away from the object until the top of the object can again be lined up in the sights. If the height of the object is H, the distance moved is m and the unit of the scale is s (10 or 12), then $H = ms$. It is quite simple to prove this by trigonometry, a subject in which the Arabs were very proficient, but it is clearly easier for the surveyor in the field to use constructive methods. Other problems solved by similar methods included finding the width of a river, or the distance between two points separated by an impassable obstruction.[8]

Methods of triangulation were unknown to the Roman agrimensores, and were introduced to Spain by astrolabic treatises such
as that of Ibn al-Saffār. Although such methods were often omitted
when the treatises were translated into Latin, an exception is the
Geometria incerti auctoris, a compilation of Hispano-Arabic inspiration which Millás Vallicrosa relates to the Arabised scientific corpus
of the Monastery of Ripoll. Roman methods existed side by side, in
Spain, with triangulation surveying. Simplified Roman procedures
seem to have been used by individual farmers, while triangulation was
carried out by professional surveyors — Christian and Muslim —
employed by large landowners. In the later Middle Ages, triangulation must have become a more common procedure.[9]

Measurement

Ropes or cords, with intermediate dimensions marked by knots, were
the usual method of measuring distances. Units of length varied
widely from one region to another, and even from town to town,
although measures were largely standardised in the Roman Empire.
Variations were not very important, provided the measures were kept
constant in a particular area. In Seville in the eleventh century tem-
plates for checking the sizes of bricks, timber beams and other
building members were kept in the mosque and used by the building
inspectors to check the work of contractors.[10] Ropes were used for
geodetic surveys as well as constructional work. In the ninth century a
survey was carried out in Iraq to measure the diameter of the earth.
This was done by measuring along a meridian, with ropes, through
one degree of latitude.[11] According to Hero, when ropes were to be
used for measuring long distances, they were to be smeared with wax
after having been tautened twice.

The hodometer, or waywiser, is a mechanical device for measuring
distances. Nowadays every motor car is equipped with such a
mechanism, but in earlier times they were quite rare, being used as a
special tool by surveyors and the military. Both Vitruvius and Hero
describe waywisers. The description by Vitruvius has caused some
problems to scholars, including Leonardo da Vinci, and it has even
been suggested that it was simply a 'design on paper'. Vitruvius is
sometimes rather shaky when he embarks on theoretical explanations, but he was essentially practical and it is doubtful whether he
would have described a machine that would not work. The main
difficulty for commentators has been that Vitruvius specifies a gear-
wheel with 400 teeth, and it was felt that such a wheel could not have

been made at the time. Andre Sleeswyk has, however, argued convincingly for the practicability of Vitruvius' design, and has made a
satisfactory quarter-scale working model which does not depart from
Vitruvius' instructions.[12] The full-scale original was a vehicle with
wheels four feet in diameter, giving a length for one rotation of about
$12\frac{1}{2}$ feet. At every rotation a single tooth on the axle moved a vertical
gear-wheel, with 400 teeth, by the distance of one tooth-width. One
complete rotation of the vertical wheel was therefore equal to 5,000
feet, or one Roman mile. When this point was reached, a protruding
tooth on the vertical wheel moved a horizontal gear-wheel by the
width of one of its teeth. This operated a ball-release mechanism, and
a ball was discharged into a metal container. At the end of the
journey, the distance could be obtained by counting the balls. The
main use of the waywiser was probably for placing milestones along
the main roads.

Astro-surveying: Determination of the *Qibla*

One of the most important problems for the scientists of medieval
Islam was the determination of the bearing of Mecca from any given
locality, this bearing being known as the *qibla* of the locality. It is
incumbent upon Muslims to face towards Mecca when they pray, and
the *mihrāb*, or prayer-niche of a mosque should be so placed that
when the congregation are facing it they are indeed facing Mecca.
The problem is a special case of the determination of the azimuth of
one place relative to another, and its solution is a trigonometric
function of the local latitude, the latitude of Mecca and the longitudinal difference from Mecca. Medieval longitude determinations,
based either on simultaneous observations of lunar eclipses in
different localities or on measuring distances between the localities,
were generally not very accurate. Latitude determinations, on the
other hand, based upon observations of the solar meridian altitude,
were usually more accurate. Inexactitudes in the geographical data
are one explanation of the incorrect orientation of some medieval
mosques. Another reason is that the qiblas of some mosques may be
incorrectly aligned is that they were not computed from geographical
data at all, but were determined by tradition. Thus, for example,
mosques in North Africa and the Indian subcontinent generally face
due east or due west, respectively. Nevertheless, the gradual refinement of the methods, until a quite remarkable degree of accuracy was

reached, represents the general advance of medieval Islamic scientists in the field of spherical astronomy.

Assuming that the three requisite quantities are known, the problem can be solved by applying the spherical cotangent rule, which is a formula combining trigonometric functions of the three quantities expressed as spherical angles. Some Muslim astronomers contented themselves with approximate solutions, which were adequate for practical purposes, but the exact solutions proposed by certain astronomers were equivalent to the cotangent rule, although they were expressed in different terms. Among those who proposed correct solutions were Habash al-Hāsib (*fl.* Baghdad and Damascus *c.* 850), al-Nayrīzī (*fl.* Baghdad *c.* 900) and al-Bīrūnī (d. *c.* 1050). The culmination of Muslim efforts in this field is represented by the qibla tables of al-Khalīlī (*fl.* Damascus *c.* 1365). His tables, based upon an accurate formula, listed the qiblas for each degree of latitude from 10 to 56, and each degree of longitude difference from Mecca from 1 to 60. Al-Khalīlī's tables thus contain a total of almost 3,000 entries and the qibla is computed to degrees and minutes. The vast majority of the entries are either correct, or are in error by one or two minutes, a remarkable achievement.[13]

Notes

1. L. Sprague de Camp, *Ancient Engineers* (Tandem Books, London, 1977), p. 34.
2. O.A.W. Dilke, *The Roman Land Surveyors* (David and Charles, Newton Abbot, England, 1971), p. 23.
3. Ibid., p. 34.
4. Vitruvius, *De Architectura*, ed. with Eng. trans. by F. Granger (2 vols. Loeb Classics, London 1931, this reprint 1970), vol. 2, bk 8, Ch. 5, pp. 179-81.
5. Claude Cahen, 'Le Service de l'irrigation en Iraq au début du XIe siècle', *Bulletin d'études orientales*, vol. 13 (1949-51), pp. 117-43 (Arabic with Fr. trans.).
6. Dilke, *Roman Land,* pp. 37-40.
7. Ibid., p. 70.
8. E. Wiedemann, *Aufsätze zur arabischen Wissenschaftsgeschichte* (2 vols., Olms, Hildesheim, 1970), vol. 1, pp. 577-96.
9. Thomas F. Glick, *Islamic and Christian Spain in the Early Middle Ages* (Princeton University Press, Princeton, 1979), pp. 228-9.
10. E. Levi-Provencal, *Trois Traités Hispaniques de Ḥisba* (Arabic text — Cairo, 1955), p. 34.
11. The Banū (sons of) Mūsà, *The Book of Ingenious Devices*, trans. and annotated by Donald R. Hill (Reidel, Dordrecht, 1979), p. 4.
12. Andre W. Sleeswyk, 'Vitruvius' Waywiser', *Archives Internationales D'Histoire des Sciences*, vol. 29 (1979), pp. 11-22. See Vitruvius, vol. 2, bk 10, Ch. 9, pp. 319-25.
13. D.A. King, 'Kibla' in EI, vol. 5, pp. 83-8.

PART TWO:
MECHANICAL ENGINEERING

8 Water-raising Machines

Introduction

Water-raising machines were (and are) used for a number of purposes, of which the most important was that of crop irrigation. They were also for supplying water for private and communal purposes, for pumping flood water from mines, bilge water from ships, and, according to Hero of Alexandria (*fl.* mid-first century AD), for fire fighting. Our information on these machines comes from both archaeological and written sources. Also, since several types are still in use today, their operation can be understood by examining working machines. A number of designs, practicable and otherwise, are described in Arabic machine treatises, notably that of al-Jazarī, composed in AD 1206.

Water-raising is obviously one of the most important activities for any society, but more especially for those in climates with insufficient rainfall, where the maintenance of an adequate water supply from wells or streams is literally a matter of life or death. We have already seen in Chapter 2 that irrigation and water supply was often possible by means of channelling water under gravity through canals, aqueducts and qanats. There were many cases, however, for instance where wells were the only source of water, which made some method of water-raising a necessity. Where watercourses were well below the level of the neighbouring fields and communities some kind of high-lift machine was needed. But even in low-lying irrigated regions such as Egypt and lower Iraq, some means of raising water over the banks of canals and rivers had to be used. Water was also often raised to the upper storeys of large buildings.

The history of each type of machine, as far as this can be ascertained, is discussed below for each individual type, but certain general observations can be applied to the whole array of machines. It is worthy of note, in the first place, that none of the more complex machines had been invented before the classical period. Alone of all the machines that will be described, only the simple shaduf or swape

127

was known in Antiquity. The earliest method of raising water from a well was almost by means of a bucket lowered down to the water on the end of a rope. The windlass was a development from this system. A wooden cylinder is installed horizontally over the well, its end rotating in bearings supported by columns. The rope to which the bucket is attached is wound around the cylinder, which is rotated to lower and raise the bucket. The windlass is still in use today in areas where wells are an important source of water, but it is usually turned by a crank, a component that did not come into everyday use until the Middle Ages. In earlier times the cylinder was turned by means of a large spoked wheel attached to one of its ends.[1] Another early method was to use a camel or an ox running down a slope away from the well to draw up the bucket. It is quite simple to devise a means for automatically unloading the bucket at the mouth of the well into a channel leading to the field.[2]

There are no real problems in describing the construction of the more sophisticated machines: most of them have been described in the works of classical and medieval writers, and several types can still be seen serving their purposes today. It is rather more difficult to assess their capabilities, since there does not appear to have been any systematic study of their characteristics in terms of power, efficiency, discharge rate, and so on. The early writers, in rare comments on such matters, are unreliable. This is no fault of theirs, but is due to our inability to assign exact modern quantities to the measures that they used. Some attempt has to be made, however, since a knowledge of a given machine's characteristics can help to explain why it was chosen to meet a certain set of needs.

There can be no certain answers about the history of the origins and diffusion of these machines, although some credible conjectures may be made. Some of the reasons for this uncertainty are already familiar to us — almost total lack of written records from AD 450 to 700, scarcity of information about Byzantium and Sasānid Iran. Almost all the machines were described by Vitruvius in the first century BC, but this does not mean, of course, that they were invented at about that time. Nor does it mean, as some writers have assumed, that they were Roman inventions. To complete the list of reservation, there is a totally confused nomenclature in our sources, both in Greek papyri and in Arabic writings.[3] One can never be sure that a particular machine is being referred to from the name alone; some other information about hydraulic conditions or the kind of motive power used is always necessary for identification. The main result of this confusion is that any assessment of the relative popularity of the various machines must be conjectural. It is unfortunate, in this respect, that modern writers have not always differentiated clearly between the two most important machines — the chain-of-pots driven through gears by animal power, and the gearless wheel driven by water. Admittedly, a wide variety of names are used for these machines by Arabic writers, but we do find some consistency in Syrian usage, where the former is almost always called the saqiya and the latter the noria (*na'ūra*). This distinction is used throughout the present work. One word *hannāna* — seems to have been reserved for the noria. It refers to the 'organ music' produced by the rotation of the wheel and although it is fairly rare in occurrence its presence in a text allows one to assume with some confidence that the machine in question is a noria.

Water-raising machines are complementary to water supply and irrigation, and so the observations made in Chapter 2 about the interrelationships between the needs of society and the support of technology apply also to the use of these machines. The period between the end of the Roman Empire and the Renaissance, in Europe, shows a marked decline in standards of water supply and sanitation, and our attention is therefore inevitably drawn to the situation in the classical world and in Islam. The role of water-raising machines in providing for the needs of society is clearly of importance to social and economic historians, but they are also of considerable significance in the history of mechanical technology, since many of the ideas and components incorporated in them were to enter the vocabulary of European engineering at a later date. We shall consider first the machines in common use and then some of the designs described in al-Jazarī's machine book. The machines that are of importance in western Asia and Europe are the following:

1. The *shadūf* or swape
2. The screw
3. The tympanum
4. The *sāqiya* — the chain of pots
5. The noria — a wheel driven by water
6. Pumps

Traditional Machines

The Shādūf

Sometimes called the swape, this is probably the earliest machine used by man for irrigation and water supply. It is illustrated as early as 2500 BC in Akkadian reliefs and about 2000 BC in Egypt.[4] It has remained in use to the present day and its application is world-wide, so that it is one of the most successful machines ever invented. Its success is probably due to its simplicity, since it can easily be constructed by the village carpenter using local materials. For fairly low lifts it delivers substantial quantities of water. It consists of a long wooden pole suspended at a fulcrum to a wooden beam supported by columns of wood, stone or brick. At the end of the short arm of the lever is a counterweight made of stone or, in alluvial areas where stone is not available, of clay. The bucket is suspended to the other end by a rope (Figure 8.1). The operator lowers the bucket into the water and allows it to fill. It is then raised by the action of the counterweight and its contents are discharged into an irrigation ditch or a head tank. When the water has to be raised through a greater height than can be managed by one shaduf, then a number are used in series.

Figure 8.1: Shaduf

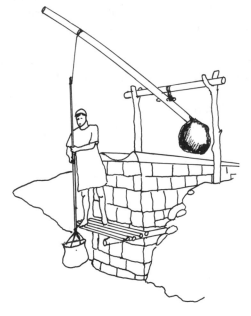

A ramp is excavated up the tank, with small platforms at intermediate levels to provide firm standing for the operators. The first shaduf lifts water from the source and its bucket discharges into that of the second machine, and so on.[5]

A variant of the shaduf is called the *dāliya* by the eleventh-century Iraqi author of *Kitāb al-Hāwī*.[6] This was a larger machine, its balance arm being up to 14 metres long, and instead of the counterweight there was a robust platform made of timber planks. Five or six men stood on the platform to raise the bucket, which was filled and discharged by another man, who presumably used some kind of tripping device to tilt the heavy bucket at the discharge point. The men stepped off the platform to allow the bucket to descend. This implies that they had to climb to the top of a bank to remount the platform after each discharge, which would have been a tiring business. There is a passage in another Arabic treatise, however, that shows one way of lessening the effort.[7] There were two daliyas in parallel and two teams of men, so that a team would have had more time to climb to the upper level than was the case with a single machine. The output was, of course, considerably higher. A second variant on the shaduf was the flume-beam swape. In this machine the pole is replaced by a wooden channel, having a large scoop opening into the channel at one end and a counterweight at the other. When the scoop is full it is raised by the counterweight and its contents run through the channel into the discharge point. Although this machine was used as a component in three of the devices described by al-Jazarī, it was never widely used in the Middle East. It is, however fairly popular in India, where it probably originated.[8]

The Screw

This machine, also known as the water-snail, was described by Vitruvius in the first century BC, with clear instructions for its manufacture.[9] The rotor consisted of a wooden cylinder, its length/diameter ratio 16:1 (Vitruvius says 'as many inches thick as it is feet long'). At either end the circumference was divided into eight equal arcs, and these points on the circumference were joined by parallel lines along the length of the rotor. The length was divided into sections, each equal to one eighth of the circumference, and a ring was drawn around the rotor at each division. The surface was thus divided into a pattern of small squares. The blades were constructed by taking a flat strip of willow or osier, coated with liquid pitch, and placing its end on one of the divisions of the circumference at the end of the shaft. Then it was drawn obliquely around the rotor shaft and nailed on to it at the intersection of the two lines, one eighth of the circumference along and at the same distance around the shaft — the pitch was therefore 45 degrees (see Figure 8.2). The winding was continued in this way, the spiral making just over five circuits of the rotor by the time it reached the other end. Other strips were nailed to the first until a laminated blade had been built up to twice the diameter of the shaft. A further seven blades were formed in the same way. A cylindrical wooden case was made to fit around the blades, constructed like a barrel, the planks painted with pitch and bound with iron hoops. The description of the bearings is somewhat obscure, but the ends of the rotor were probably provided with iron caps from which iron spigots protruded. The spigots rotated in iron journals fixed to the supports.

The method of rotating the screw is also obscure: Vitruvius simply says 'And so the screws are rotated by men treading — *et ita cocleae hominibus calcantibus faciunt versationes*'. It is clear that a treadmill was used and it is possible that power was transmitted from this to the screw through a pair of gear-wheels, since these were already in use in mills. We cannot be sure. Medieval commentators on Vitruvius show the screw turned by a crank, but this is certainly an anachronism. Vitruvius recommends that the screw be installed at an angle of about 37 degrees, which would give a high lift with a low output. Remains of Roman screws and their mountings found in Spanish mines indicate an angle of about 15 degrees. The angle chosen would depend upon the relative importance of lift and output.[10]

The invention of the screw is ascribed to Archimedes by ancient authors, and of course the device usually bears his name. In his

Figure 8.2: Screw

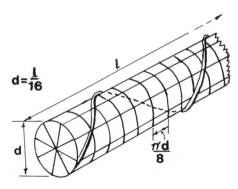

thorough study of Greek and Roman mechanics,[11] Drachmann mentions that there has been a tendency on the part of some modern writers to doubt this ascription, but he is convinced that the doubts have no foundation. He makes the point, however, that the screws found in Spanish mines show a single spiral of copper instead of the eight wooden windings described by Vitruvius, and asks why Vitruvius described this design when a single spiral is sufficient and takes up less space. His answer is that this was the original Archimedean pattern and that the single spiral was introduced later, probably by the engineers in the copper mines. He also suggests that Archimedes, who is said to have invented the screw during a stay in Egypt, saw the drum or tympanum (see below) in everyday use there and applied his knowledge of spirals to derive the screw from the drum. And since the tympanum has eight compartments, so did the original screw. Drachmann concludes: 'The invention was a stroke of genius, and it comes almost as a relief to find that after all it was not made out of the whole cloth, so to speak. I have always felt that this was rather too much!' This explanation is perfectly credible and accords well with documented examples of the process of invention. The screw was in use in Egypt in Ptolemaic times, and spread from there to North Africa, Spain and France during the Roman Empire. Its main use was for irrigation, but it was also used for drainage in mines and for extracting water from the bilges of ships.[12] According to Forbes it was still in common use in Upper Egypt and other parts of the Arab world in 1965, but had disappeared from the Delta region.[13] The screw is very rarely mentioned by Arabic writers.

The Typanum or Drum

As described by Vitruvius,[14] this machine consisted of a wooden axle with iron pegs protruding from its ends. The pegs were housed in iron journals that were supported on stanchions. Two large timber discs made up from planks were mounted on the axle, and the space between them was divided into eight segments by wooden boards. The perimeter was closed by wooden boards, there being a slot in each segment to receive the water. Circular holes were cut around the axle in one face of the drum, one hole to each segment. The whole machine was coated with tar. The water discharged into a small tank connected to a channel through which the water flowed to the fields or to a drainage area. The wheel was operated 'by men treading'. Vitruvius does not say whether this was done by fixing cleats to the rim of the tympanum or by means of a separate treadmill on the same

Figure 8.3: Tympanum

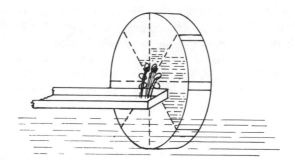

axle. The machine was limited to fairly low lifts. In order to raise a significant quantity of water it must dip into the source for about one third of its diameter and, as indicated in Figure 8.3, this implies that the height of lift was also about one third of the diameter, for example a lift of 1 metre for a wheel 3 metres in diameter. On the other hand, it was suitable for use in restricted spaces such as mine adits.

A tympanum in reverse is described in an Arabic version of the *Pneumatics* of Philo, who flourished about 230 BC.[15] The author says that this is similar to 'the drums used for irrigation'. In this case the water drives the wheel, by running into and being discharged from the slots in the perimeter. It was intended to cause a whistling sound by some pneumatic means not clearly described in the text. It might therefore be assumed that the invention of the tympanum was earlier than the time of Philo, were it not for the fact that the part of the manuscript containing this passage cannot be confidently attributed to Philo. All the 'Philonic' corpus of manuscripts in Arabic contains Islamic additions, and this passage may be one of these. On the other hand, Vitruvius makes it clear in his description of the various types of water-raising machines, that he was not dealing with new inventions, but with machines that had been in use for some time. We have already discussed Drachmann's suggestion that Archimedes modelled the screw on the drums that he had seen working in Egypt. If this is correct, then the machine was already established by the middle of the third century BC, in which case the passage in the Arabic text describing a device derived from the drum could safely be attributed to Philo. The evidence seems to be in favour of an Egyptian origin for the machine in the first half of the third century BC. There are very few definite indications for the use of the drum, either in the classical world or in Islam, although Strabo mentions the use of wheels driven by treadmills in Egypt.[16]

The Saqiya

The word saqiya is used here to denote the chain-of-pots driven through a pair of gear-wheels by one or two animals harnessed to a draw-bar and walking around a circular track. The chain-of-pots could also be driven by a treadmill, mounted on the same axle as the wheel carrying the chain-of-pots. This method was in use in Roman times and is also described in an Arabic manuscript dealing with the water-raising methods of the people of Isfahan.[17] The animal-driven saqiya is the most effective of the traditional machines for high-lift duties and its most common use is for drawing water from wells. Thorkild Schiøler has made an exhaustive study of the saqiya, and his published results [18] are based upon archaeological and literary sources, together with personal visits to both working and ruined water-raising installations. I have drawn heavily upon Schiøler's book for the constructional details and history of the machine, which can be made in a variety of ways. There is no better way of giving the basic constructional details of the machine than to quote Schiøler's description of a saqiya in Ibiza. It was still in use in 1955 but was in ruins a few years later.

> The crux of these machines is the gear, which has a single function — that of altering the motion from horizontal to vertical. The gear itself consists of a lantern pinion and a large cog-wheel. In the following, the latter will be referred to as the potgarland wheel, since it also operates the paternoster chain.
>
> The most convenient way of describing such a machine is to follow the power transfer through the machine, starting with the draught animal and ending where the water leaves the machine. The draught animal is usually a mule or a donkey wearing on its shoulders and neck a collar harness that transmits the power through two traces to a double-tree fastened to the drawbar. The drawbar passes through a hole in the upright shaft. The leading rein, which is attached to a second bar protruding from the shaft, forces the animal to follow the circular track, although it cannot see this because its eyes are blinkered.
>
> The lantern pinion is fixed to the upright shaft by means of two sets of spokes, one for the upper rim and one for the lower. The spokes are made in one piece from bars equal in length to the outer

diameter of the rims and countersunk inside the shaft so that they bisect each other in the same plane. To achieve this, the hole through which one bar passes through the shaft is made twice the height of the hole for the other bar. When the spokes are in place, the larger hole is wedged to keep the spokes locked in position.

Both wheel-trims of the lantern pinion are made by the carpenter from six or seven peices of wood, cut to a template ... The pins of the lantern pinion are slightly conical in shape and are of a soft wood because they are easy to make and the peasant puts in new pins approximately every other year. The cogs with which the pins mesh are difficult to make so a hard wood is used that will last for many years.

Before we leave the lantern pinion there are a few other details concerning the upright shaft that are of interest. At its base, the shaft ends in an iron gudgeon, which is bedded in a stone. The end of the shaft is provided with an iron ring to prevent it from splitting. The upper bearing for the shaft is located about 20 centimetres above the lantern pinion and consists of a beam with a yoke or liner fixed in position by two wooden spikes. [The beam is supported on two plinths of soft volcanic rock.] (See Figure 8.4.)

The potgarland wheel has two functions: it is primarily a cogwheel, but it also forms the drum that carries the potgarland. Each cog and carrying peg are made in one. The potgarland wheel of the Spanish noria has only one rim, which is made from six pieces, and four spokes made from two continuous bars equal in length to the diameter of the wheel. The short shaft is very thick and is provided with a gudgeon at each end, turning in wooden blocks.

In order to prevent the wheel from going into reverse, the machine is provided with a pawl mechanism, which acts on the cogs of the potgarland wheel. To appreciate the vital function of this pawl it is only necessary to remember that the draught animal is subjected to a constant pull both when moving and when standing still. This pull is exerted by the part of the potgarland carrying the full pots. The pawl is activated in two cases — when the animal is to be unharnessed and in the event of the harness or traces breaking. Without the pawl the machine would turn backwards at great speed and, after one revolution, the drawbar would hit the animal on the head. At the same time, many of the pins of the lantern pinion would break and the pots smash.[19]

Figure 8.4: Saqiya

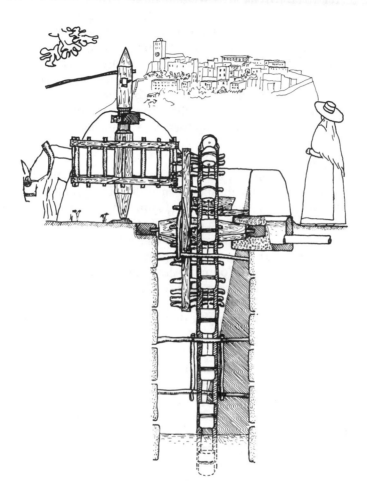

Some other points about the construction and performance of the saqiya are made by Schiøler. The potgarland is provided with two guide rods which keep the potgarland on the right course up to the potgarland wheel. These two rods are placed on either side of the potgarland and at an angle of about 45 degrees to it. Without the rods, the potgarland would slide out along the carrier-pegs and finally fall out of them.[20] On the construction of the machine in general, Schiøler says that it is a complicated and sensitive piece of machinery consisting of more than 200 different parts, and that the shaduf is simpler and therefore more reliable.[21] We shall return to this point later, but should mention that the shaduf, for all its advantages, cannot raise water from deep wells. Finally, there is a matter raised by Landels with reference to the possibility of the potgarland slipping due to the out-of-balance load between the full and empty pots.[22] Schiøler states, correctly, that the possibility of slip depends upon three factors: the pulls exerted by the full and empty pots and the coefficient of friction between the ropes and the pegs of the potgarland wheel. He found that there was no tendency to slip with ropes and earthenware pots, but that there was such a tendency when the pots were replaced by buckets.[23]

One of the problems in water-raising engineering is that of raising large quantities of water through a small lift. The problem can be solved by using a spiral scoop-wheel (Figure 8.5), which raises water to the ground level with a high degree of efficiency. This machine is very popular in Egypt nowadays, and engineers at a research laboratory near Cairo have been trying to improve the shape of the scoop in order to achieve maximum output. Although it appears very modern in design, this is not the case, since a twelfth-century miniature from Baghdad shows a spiral scoop-wheel driven by two oxen. The transmission of power is the same as that employed with the standard saqiya.[24]

The chain-of-pots is another of the machines mentioned by Vitruvius, although in this case it was driven by men working a treadmill — there is no mention of gears.[25] There is, however, archaeological evidence for the existence of the geared machine driven by animal power in the Roman world in the first century BC, including particularly impressive hydraulic works at Tuna al-Gabal in Egypt.[26] At first saqiyas were a luxury article and were used for water supply and irrigation on the estates of rich men. By about AD

Figure 8.5: Spiral Scoop-wheel

250 they were in use on large estates, but they probably did not come into widespread use for water supply and irrigation until the fourth or fifth century AD, with the introduction of the pawl mechanism and earthenware pots.[27] All the evidence suggests that it originated in Egypt in the third century BC and spread rapidly over the Middle East.[28] As is the case with other machines and techniques, there is a gap in our knowledge of the saqiya from the fifth century to the advent of Islam, but we must assume that it continued in use during this period. There are unambiguous references to the saqiya in the works of Arab geographers of the tenth century,[29] and there is a full description of the machine in a 17-volume Arabic encyclopaedia written by Ibn Sīda, who died in 1066.[30] In the agricultural manual written by the Spanish Muslim Ibn al-'Awwām in the twelfth century there is another description, which contains the interesting comment that the potgarland wheel should be made heavier than is usual in order to make the machine operate more smoothly. This is a clear indication that Ibn al-'Awwām understood the principle of the fly-wheel.[31] The use of the saqiya was introduced to the Iberian peninsula by the Muslims, where it was massively exploited. Such diffusions continued: not only did Christian Spanish engineers take it to the New World, but it was also in use in the rest of Europe and in many places remained in use well into this century.[32] There can be no doubt as to the importance of the machine for water supply and irrigation.

The Noria

The noria is perhaps the most significant of the traditional water-raising machines. Being driven by water, it is self-acting and requires the presence of neither man nor animal for its operation. Figure 8.6 shows a model of one of the great wheels on the Orontes at Hama in Syria, and demonstrates the complexity of the design. The construction is as follows: first, four continuous beams constituting the four double spokes are mounted on either side of a frame wedged tightly in position round the hub. Next, the inner rim is constructed and, at the same time the fan-shaped spokes are nailed in place. The wheel has to be turned at intervals during assembly to make sure that it turns in a single vertical plane. Finally the outer rim for the compartments is mounted. The design of the compartments is shown in Figure 8.7.[33] Instead of compartments, norias may have earthenware pots, similar to those of the saqiya, lashed to the rim. The largest of the Hama wheels has 120 compartments; the description of a noria in the *Kitāb al-Hāwī* specifies 80 pots each of about 7.65 litres.[34] The

Hama wheels discharge into the end of an aqueduct that carries the water to the town and the fields. A similar system at Toledo was described by al-Idrîs in 1154.[35] The largest of the Hama wheels have a diameter of about 20 metres and are an impressive sight, but al-Idrīsī says that the lift of the wheel at Toledo was 90 cubits or about 50 metres.

The origin of the noria, as with the other water-raising machines, is somewhat obscure. It is mentioned, clearly but briefly, by Vitruvius and was therefore in use in the first century BC.[36] Needham suggests that it was invented in India and reached the Hellenistic world in the first century BC and China in the second century AD.[37] This is a possibility, but it is usually the case that the introduction of a new machine antedates the first literary reference by a considerable period, and the noria may therefore have already been established in the Hellenistic world before the time of Vitruvius.

The noria can only be used in running water, and it is also most suitable for use when it is required to raise the water some distance

Figure 8.6: Noria

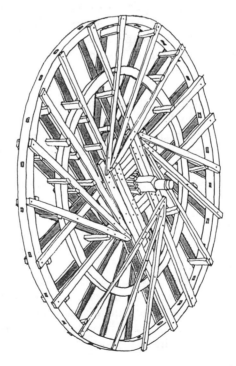

Figure 8.7: Detail of Noria

from the level of the stream. These two conditions are common in Syria and Iran, and it is suggested that the machine originated in one of those countries. The earliest reference we have for use of the noria in Islam occurs in one of the historical works of al-Balādhurī (d. 892). He tells us that Bilāl, the son of a certain Abu Burda, excavated a canal in the Basra region and erected norias (*hannāna*) along its banks. This event probably occurred in the second half of the seventh century, since Bilāl's father died in 722 at an advanced age. When Bilāl had constructed the machines he moved the market to the banks of the canal. This suggests that a community was not viable without some efficient means of raising water, and also implies the existence of norias in Iraq before Islam.[38] The oldest known reference to the Hama wheels occurs in the Geography of Yaqūt (1229) quoting Ahmad ibn al-Tayyib, who died in 885: 'Hama is a town with a stone wall, beside which there is a big stone building. The river flows and waters the gardens and drives the water-wheels.'[39]

Al-Muqaddasī, the great Arabic writer on geography, in a work composed between 985 and 990, seems to have been particularly impressed by the norias in Iran. He mentions the many norias on the river at Ahwāz in Khuzistan,[40] and the norias installed below the dam called Band-i-Amir, built by the Buwayid Amir'Adud al-Dawla (d.983) over the river Kūr between Shiraz and Istakhr.[41] Norias were even used on non-perennial streams: al-Istakhrī (d. after 951) mentions a well in a village in the Iranian province of Fars that was dry for most of the year until a certain time 'when the water gushes from it up to ground level and the flow from it turns a noria which is used to irrigate the farms'.[42] Ibn Jubayr (d. 1217) describes a noria in the town of Ra's al-'Ayn in Upper Mesopotamia, which lifted the water from the River Khabur to the gardens.[43] These reports, which are but a selection from the many references to norias in the works of Arabic writers, bear witness to the widespread use of the noria in the world of Islam.

Pumps

The only type of pump that we know to have been in everyday use in the Roman world is the force-pump. Vitruvius describes one of these clearly, and attributes its invention to Ctesibius (third century BC).[44] A similar pump appears in the *Pneumatics* of Hero of Alexandria (*fl.* mid-first century AD), who says that it was intended for fire-fighting.[45] The pump of Ctesibius is shown in Figure 8.8. The whole assembly was made of bronze, the pistons being turned on the lathe to ensure a tight fit. The delivery pipes from the cylinders led into a vessel capped by a close fitting funnel, from the top of which the single delivery pipe was led out. The piston rods were connected by pin-joints to a rocker arm which oscillated about a central fulcrum when the operators worked the pump. The advantage of the force-pump is that it will lift water to any height consistent with the piston, cylinder, valves and delivery pipe being able to withstand the hydrostatic pressure. The notable drawbacks of the machine are that the pumping mechanism is submerged, and that if the water-level falls, the cylinder will not fill. It is clear that the Romans used piston pumps of this exact type but only in very small sizes. Of the handful of specimens which have been unearthed, at Silchester, Metz, Germain-en-Laye and Bolsena, for example, all measure only inches in both bore and stroke and apparently they were hand operated. All this suggests that their application was limited to minor tasks such as filling domestic cisterns. On the ensuing history of the piston pump, Smith writes as follows:

Figure 8.8: Force-pump

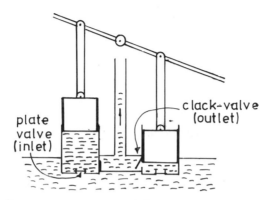

We simply do not know if piston pumps were part of Dark Age and medieval European technology. Their appearance in various fifteenth-century technical manuscripts — such as those of Francesco di Giorgio Martini, Taccola and Leonardo da Vinci — may merely be the first pictorial record of what was, in reality, a continuing tradition. If not, at least three possibilities are open to consideration: Europe inherited the idea from Islam whose engineers *were* familiar with piston pumps; the discovery of Vitruvius' manuscript in 1408 prompted fifteenth-century renderings of the Ctesibian pump; or European technology brought forth the machine as part of current developments in which the introduction of the crank and the connecting rod combination was important.[46]

The question of Islamic piston pumps will be discussed in the final part of this chapter. It is interesting to note that when the European piston pump makes its appearance in the writings of Taccola (*c.* 1450)[47] and his successors, a suction stage is already incorporated. The use of a suction pipe on the inlet side of the pump permits water to be drawn from a theoretical depth of about 33 feet, although in practice not more than 25 feet is attainable. Oddly enough, the European pumps of the fifteenth century are almost *all* suction stage — the delivery is little more than the stroke of the piston.

Table 8.1: Output of Water-raising Machines

Machine	Winter area	Summer area
Noria	350–400	80
One-ox Saqiya	70	30
Two-ox Saqiya	150	70
Daliya	80–130	53–60
Shaduf	70	30

Assessment of the Traditional Machines

Any consideration of the water-raising machines in everyday use must take account of two facts: the machines are still in use today, often being preferred to modern machines because they can be quickly repaired by the village carpenter, whereas a diesel pump has to be maintained by a skilled mechanic; in medieval times they helped to provide water for a thriving agriculture and for very large urban communities. They deserve, therefore to be considered as successful machines in their own right and not simply as a stage in the development of modern machinery. The most successful of them are the shaduf, the noria and the saqiya. If we include with the saqiya the spiral scoop-wheel, then these machines can be said to meet, taken together, the varied demands of water-raising. The figures (Table 8.1) for the areas that could be irrigated by each machine are derived from *Kitāb al-Hāwī*.[48] The areas are in *jarībs* — 1 jarib = 1,366 square metres.

These must be average figures, since the output of a machine depends upon various factors, and the amount of water needed for a given area depends upon the type of crop. Nevertheless, they can be accepted with some confidence. One notices, for example, that the summer capability of the noria is less by comparison with its winter capability than is the case with the other machines. This can be explained by the fact that, in addition to increased evaporation, only the noria is affected in summer by the decreased flow of the streams. Also, the two-ox saqiya has more than double the capability of the one-ox variety, presumably because the operation was smoother and neither of the oxen tired as quickly as an animal on its own. According to these figures, one shaduf was capable of irrigating about four hectares, the same area serviced by the one-ox saqiya. It might be thought that the simpler, cheaper shaduf would always be preferred, but of course the shaduf cannot be used to draw water from deep wells, and operating one is back-breaking labour. The question of which machine to use is affected by various factors: the nature of the source of water, the height of lift, the type of constructional materials available and the quantity of water required. For example, the noria has a relatively high output and does not require constant attendance, but it can only be used in running water and is expensive to build and maintain.

It is always a risky business to try to assess the performance of early machines from the information given in works like *Kitāb al-Hāwī*, if only because we can never be sure that we are using the correct modern equivalents for the old weights and measures. The author is, however, rather more specific than is usually the case, so it is worthwhile making some quantitative evaluations, and testing them against modern figures. The noria, says our author, irrigated one jarib in every hour of a 24-hour day, and the jarib required up to 300,000 *ratls*, or about 153,000 litres. It had 80 pots, each with a capacity of 15 ratls = 7.65 litres; at each rotation, therefore, 612 litres was delivered. The 153,000 for one jarib would cover the ground to a depth of 11.2 cm, assuming (we have no choice) that our conversion factors are correct. This is an interesting result, since it agrees well with the 'ankle-depth' that was normal for one watering.[49] But there are problems; from the size of the pots we can estimate that the wheel was about 11.5 metres in diameter, and allowing for immersion of the pots and the drop from the pots into the outlet channel, the lift may have been about 10 metres. To deliver 153,000 litres in an hour the wheel had to rotate at about 4 revolutions a minute, so the current speed had to be of the order of 3 metres a second, which is very high. At an efficiency of 33 per cent the output and input horsepowers would have been 5.58 and 16.9 respectively; again, these are very high figures. The wetted surface — i.e. the area of the blades in contact with the water at any instant — would have been about 0.44 square metres. Since this is a fairly modest area, the main obstacle to accepting the output of 153,000 litres an hour is the high current speed. We shall see in the next chapter, however, that very powerful mills driven by paddle-wheels were used on the Tigris and Euphrates, and we know that the norias on the Sarat canal were an obstacle to navigation.[50] We may infer from this that they were large and also, perhaps, that they were associated with dams to increase the current speed. Collett gives figures for the performances and sizes of modern norias in East Asia, which cover a very wide range indeed, from small wheels delivering one litre a second to a wheel of 20 metres in diameter delivering 80 litres a second, or 288,000 litres an hour.[51]

Leaving norias for the moment, and referring to figures derived by Molenaar for the average performance of the saqiya with various heights of lift, we find that this machine will lift from 8,000 litres an hour with a 9-metre height to 22,000 litres with a 5-metre height.[52] If we take the higher of these quantities and multiply it by the ratio of the highest output of the noria given in *Kitāb al-Hāwī* with the lowest output of the saqiya in the same book, we arrive at a maximum for the noria of about 293,000 litres an hour. This is not very convincing, since we have selected the figures to suit the argument for a very high output for the noria. Nevertheless, the weight of evidence is in favour of a high performance for the large noria in good hydraulic and climatic conditions against a lower, though still substantial output of the saqiya. We have already noted that the spiral scoop-wheel, which is essentially a variant of the saqiya, was in use no later than the twelfth century. Its growing popularity today is no doubt due to its high productivity for fairly small lifts. According to Molenaar, its average output varies from 36,000 litres an hour for a lift of 180 cm, to an output of 114,000 litres an hour for a 30 cm lift.

Water-raising Machines in al-Jazarī's book

Al-Jazarī completed his book on machines in AD 1206 in Diyar Bakr, at which time he had been 25 years in the service of the ruling family of Artuqid princes.[53] Most of the devices that he describes are water-clocks and various types of automata — we shall be discussing this aspect of his work in the last two chapters of this book. Because he dealt mainly with small devices, some writers have assumed that his water-raising machines were merely fanciful notions with no practical purpose.[54] I do not believe this to be the case. There was a demand from al-Jazarī's masters for devices that would provide amusement and aesthetic pleasure, but it is also highly probable that his responsibilities included the design and construction of public works. In this capacity he would have appreciated the need for improving the efficiency of water-raising methods, and have attempted to devise means to this end. Apart from their potential as practical machines, his designs have the added significance of incorporating techniques and components that are of importance in the development of machine technology.

Figure 8.9: Al-Jazari's First Water-raising Machine

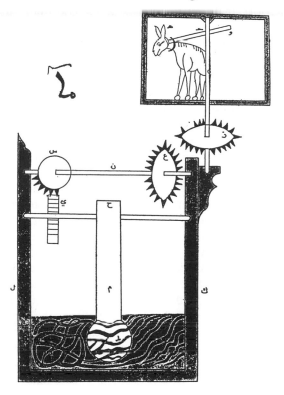

The first machine is shown in Figure 8.9. Two strong stanchions were erected in a pool, and two axles, one vertically over the other, rotated in journals housed in these stanchions. On the lower axle was a flume-beam swape with a capacity of 30 ratls or more, i.e. 15 litres upwards, and a lantern pinion. On the upper axle there were two gear-wheels, one of them having teeth on only one quarter of its perimeter, the other being a normal gear-wheel. The segmental wheel meshed with the lantern pinion, and the second with a horizontal gear-wheel, the vertical axle of which passed through the floor of the operating room. On its upper end was a draw bar, to which a donkey was tethered. As the donkey walked in a circular path, the upper horizontal gear-wheel was turned and the teeth of the segmental gear-wheel entered the bars of the lantern pinion. The scoop was therefore raised and the water ran through the channel and discharged into an irrigation channel. When the teeth disengaged from the lantern pinion the scoop fell back into the water and was refilled in time for the next rotation. The segmental gear is an interesting part of this machine. A similar wheel first appeared in Europe in Giovanni de' Dondi's astronomical clock, completed about 1365.[55] This type of gear was, however, known in Islam before the time of al-Jazarī; an eleventh-century Spanish Muslim called al-Murādī used it in some of his devices (see Chapters 11 and 12). The second of al-Jazarī's machines is a quadrupled version of the first, having four swapes, four lantern pinions and four segmental gears. All four swapes therefore rose and fell for one rotation of the vertical axle. Output was thus considerably increased — not perhaps quite fourfold, since the donkey may have walked more slowly because of the increased load.

The third machine is a miniature water-driven saqiya, erected as an attraction beside an ornamental pool. Water from the pool ran into an open tank through a pipe, then through a hole into a concealed, lower tank. It fell on to the scoops of a water-wheel, which turned a gear-wheel mounted on its axle. This gear meshed at right angles with another, whose vertical axle passed through the floor of the upper tank. Towards the top of this axle was a third gear-wheel that meshed at right angles with a fourth, upon whose axle was the potgarland wheel. Water from the pots fell into a tank between the two rims of the wheel, and ran through a channel to discharge into the pool. The vertical axle, for a short distance above the floor of the upper tank, was enclosed in a wide pipe, to the top of which a copper disc was soldered. A lightweight wooden model cow was fixed rigidly by a draw bar to the axle, its feet just clear of the surface of the disc — the model turned with the axle. The whole construction was quite small: the open structure around the upper pool was probably about 3 metres high, and the scoop-wheel was about 1.7 metres in diameter. It has been thought that this machine was simply an unrealised idea of al-Jazarī's based upon the principle of the animal-driven saqiya, whereas it is in fact a scaled down version of an utilitarian machine. One of these can be seen in working order on the River Yazid in Damascus. It was built not later than 1254 to serve the needs of a hospital and remained in constant use throughout the centuries, until it finally fell into disrepair about 1960. It has now been thoroughly restored by the staff and students of Aleppo University. It is a much larger version of the machine described by al-Jazarī, but instead of a scoop-wheel it has a large paddle wheel.[56] There is a similar machine in a manuscript attributed to Philo, but this is almost certainly an Islamic addition. Carra de Vaux published a drawing reconstructed from the text, but this is somewhat inaccurate. He shows a scoop-wheel, for example, whereas the text indicates a paddle wheel [57]

Al-Jazarī's fourth machine again has a donkey in a raised chamber tethered to a draw bar and turning a vertical axle. On this axle, below the chamber, is a gear-wheel meshing at right angles with a second wheel, mounted on a horizontal axle which has a crank fitted to it. The free end of the crank enters a slot-rod under the channel of a flume-beam swape, the scoop of which is in a pool. As the donkey walks in a circle the horizontal axle is turned by the gears, and the action of the crank in the slot-rod raises and lowers the swape. (Figure 8.10 does not show the arrangement clearly, so Figure 8.11 has been added.) This is the earliest evidence we have for the crank as part of a machine, although manually operated cranks had been in use for centuries.

The fifth machine is of considerable importance in the development of machine technology. It is a piston pump with two alternative means of propulsion. The first of these is a horizontal vaned wheel turned by a running stream. The axle of this wheel enters the machine itself direct, without any gearing. In the second version a paddle wheel was mounted on a horizontal axle over a stream. On the other

Figure 8.10: Al-Jazarī's Fourth Water-raising Machine

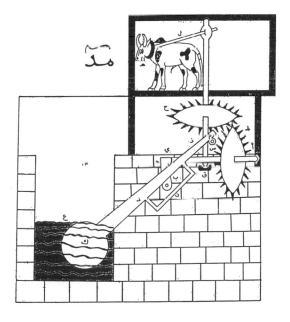

Figure 8.11: Detail of Figure 8.10

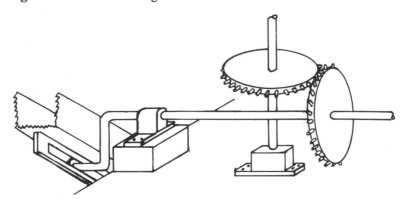

end of the axle was a gear-wheel that meshed at right angles with another gear-wheel inside the machinery box. (It is possible that the box was vertical as shown in Figure 8.12 but more likely that it was horizontal, as shown in Figure 8.13.) On the surface of the second gear-wheel, near its rim, was a vertical peg which entered a slot-rod. A long rod, soldered to the end of the slot-rod, rotated about a pivot in one corner of the box. The connecting rods were fixed to the sides of the slot-rod by staple-and-ring fittings. On the end of each connecting rod was a piston, consisting of two copper discs with a space of about 6 cm between them, the space between them filled by coiling hempen cord until the gap was filled. The cylinders, made of copper, were provided with suction and delivery pipes. All the pipes were fitted with non-return clack-valves. The two delivery pipes were joined together above the machine into a single delivery pipe. The action was as follows: when the paddle wheel turned it caused the gear-wheel on its axle to turn, and this rotated the gear-wheel in the box and the peg made the slot-rod oscillate from side to side. When one piston was on its delivery stroke the other was on its suction stroke. It was therefore a double-acting piston pump with true suction pipes, and an effective means of converting rotary into reciprocating action. Al-Jazarī says that this machine is a large version of the machines used for discharging Greek fire, thus giving us our first clear information about the design of the so-called 'Byzantine siphon'.

We are told that the height of the delivery point above the machine was 20 cubits, i.e. about 13.6 metres, and that the water issued with great force. The pistons entered the cylinders in a sliding fit — the

Figure 8.12: Al-Jazarī's Double-acting Pump

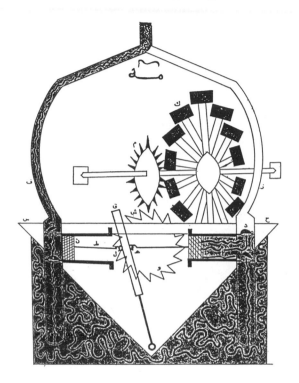

hemp ensured that leakage was not excessive. There can be no doubt
that this pump was a practical machine. A quarter-scale working
model of it was made for the 1976 World of Islam Festival in the
Science Museum, London. The construction was the same, in every
respect, as that of the pump described by al-Jazarī, except that the
drive was electric. The transmission is very smooth, and the pump
discharges a steady stream of water from the delivery pipe. This pump
embodies mechanisms from two machines — the paddle wheel from
the noria and the pumping machinery from the Byzantine siphon. Al-
Jazarī was trying to solve a specific problem, that of raising water as
cheaply and efficiently as possible from the deeply indented rivers of
Upper Mesopotamia to the surrounding fields and communities. We
cannot be sure that, for a lift of 13.6 metres, his pump was more
efficient than a noria of the same diameter, but it does seem likely that
friction losses were significantly less. It would also have been cheaper

Figure 8.13: Detail of Figure 8.12

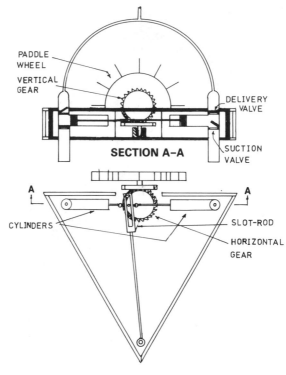

SECTION A-A

PLAN
(lid of box and delivery pipes omitted)

in materials and in labour. Unfortunately, we have no evidence for the adoption of this design in Islam except for the description of a similar, slightly improved design in a treatise written by Taqi al-Din about 1552.[58] Al-Jazarī's pump is certainly more advanced than the suction pumps that appeared in Europe in the fifteenth century, since it is self-acting, and incorporates both a suction and a delivery lift, and the double-acting principle. The possibility of its transmission to Europe, however, must remain an open question.

Notes

1. L. Sprague de Camp, *Ancient Engineers* (Tandem Books, London, 1977), p.74.

2. R.J. Forbes, *Studies in Ancient Technology*. 1st edn (6 vols., Brill, Leiden, 1955-8; only the 2nd edn of vols. 1 and 2 are used in the present work — vol. 1, 1964, vol. 2, 1965, both publ. Brill, Leiden), vol.2, p. 33.

3. Thorkild Schiøler, *Roman and Islamic Water-Lifting Wheels* (Odense University Press, 1973), p. 40 for Arabic nomenclature, p. 110 for Greek.

4. Norman A.F. Smith, *Man and Water* (Peter Davies, London, 1976), p. 7.

5. Eilhard Wiedemann and Fritz Hauser, 'Über Vorrichtungen zum Heben von Wasser in der islamischen Welt', *Jahrbuch des Vereins Deutscher Ingenieure*, vol. 8 (1918), pp. 121-54, p. 127.

6. Claude Cahen, 'Le Service de l'irrigation en Iraq au début du XIe siècle', *Bulletin d'études orientales*, vol. 13 (1949-51), pp. 117-43, p. 118 in Arabic, p. 130 in French translation.

7. Wiedemann and Hauser, 'Über Vorrichtungen', p.150. In the Oxford MS no. 954, this is the fifth chapter in the section on the machines of the people of Isfahan.

8. Joseph Needham, *Science and Civilisation in China* (Cambridge University Press, 5 vols. to date, continuing, 1954 onwards), vol. 4, pt. 2 (1965), p. 334.

9. Vitruvius, *De Architectura* (2 vols., vol. 1 1st edn 1931, vol. 2 1934, third reprint of both vols. 1970, Leob Classics), ed. with Latin text and English trans. by Frank Granger, vol. 2, pp. 307ff.

10. J.G. Landels, *Engineering in the Ancient World* (Chatto & Windus, London, 1978), p. 61.

11. A.G. Drachmann, *The Mechanical Technology of Greek and Roman Antiquity* (University of Wisconsin Press, Madison, Wisconsin and Hafner, London, 1963), p. 154.

12. Landels, *Engineering*, p. 66.

13. Forbes, *Ancient Technology*, vol. 2, p. 40.

14. Vitruvius, vol. 2, pp. 303-5.

15. Carra de Vaux, 'Le Livre des Appareils Pneumatiques par Philon de Byzance', *Paris Académie des Inscriptions et Belles Lettres*, vol. 38 (1903), pt. 1, pp. 27-235, pp. 112-14 in Arabic, pp. 201-4 in French translation.

16. Forbes, *Ancient Technology*, vol. 2, p. 16.

17. Schiøler, *Water-Lifting Wheels*, pp. 133, 136; Wiedemann and Hauser, 'Über Vorrichtungen,' p. 154; Chapter 4 of 'People of Isfahan' in Oxford MS no. 954.

18. Schiøler, *Water-Lifting Wheels*, *passim*.

19. Ibid., pp. 18-20.

20. Ibid., p. 21.

21. Ibid., p. 169.

22. Landels, *Engineering* pp. 71-2.

23. Schiøler, *Water-Lifting Wheels*, p. 22.

24. Ibid., p.79.

25. Vitruvius, vol. 2, p. 305.

26. Schiøler, *Water-Lifting Wheels*, pp. 141-8.

27. Ibid., p. 169.

28. Needham, *Science and Civilisation*, pp. 352ff.

29. Al-Muqaddasi, *Ahsān al-taqāsīm fī ma'rifat al-aqālīm*, ed. M.J. de Goeje (vol. 3 of *Biblioteca Geographorum Arabicorum* (BGA), Brill, Leiden, 1906), p. 208; and Ibn Hawqal, *Kitāb Sūrat al-Ard*, ed. J.H. Kramers, 2nd edn (vol. 2 of BGA, Brill, Leiden, 1938), p. 324.

30. Wiedemann and Hauser, 'Über Vorrichtungen', p. 129.
31. Schiøler, *Water-Lifting Wheels*, pp. 30ff.
32. Smith, *Man and Water*, p. 20.
33. Schiøler, *Water-Lifting Wheels*, pp. 37-8.
34. Cahen, 'Irrigation', p. 118 in Arabic, 130 in French.
35. Al-Idrīsī, *Description de l'Afrique et de l'Espagne*, Arabic text ed. with French trans. by R. Dozy and M.J. de Goeje (Brill, Leiden, 1866), p. 187 in Arabic, p. 228 in French.
36. Vitruvius, vol. 2, p. 305.
37. Needham, *Science and Civilisation*, vol. 4, pt. 2, pp. 361-2.
38. Al-Balādhurī, *Liber Expugnationatis Regionum*, Arabic text with critical apparatus in Latin, ed. M.J. de Goeje (Brill, Leiden, 1866), p. 363.
39. Eilhard Wiedmann, *Aufsätze zur arabischen Wissenschaftsgeschichte* (2 vols., Olms, Hildesheim, 1970), vol. 1, p. 223.
40. Al-Muqaddasī, p. 411.
41. Ibid., p. 444.
42. Al-Istakhrī, *Kitāb al-masālik wa'l-mamālik*, ed. M.G. 'Abd al-'Āl al-Hīnī (Cairo, 1961), p. 150.
43. Ibn Jubayr, *Rihla*, ed. W. Wright, Leiden 1852, 2nd edn of Wright's text, amended by M.J. de Goeje (Brill, Leiden, 1907), p. 243.
44. Vitruvius, vol 2, p. 311.
45. Hero of Alexandria, *Pneumatics*, ed. B. Woodcroft (1st edn 1851, facsimile edition with introduction by Marie Boas Hall, London and New York, 1971), p. 44.
46. Smith, *Man and Water*, pp.97-9.
47. Mariano Taccola, *De Ingeneis*, Frank D. Prager and Gustina Scaglia (eds.) (MIT Press, Cambridge, Mass. and London, 1972), pp. 46 and 50-1.
48. Cahen, 'Irrigation', pp. 118-30.
49. Thomas F. Glick, *Islamic and Christian Spain in the Early Middle Ages* (Princeton University Press, Princeton, 1979), p. 71.
50. Ibn Hawqal, *Kitāb Sūrat*, p. 242.
51. John Collett, 'Water Powered Water Lifting Devices', prepared for the Land and Water Development Division FAO, 1980. (Mr Collett was kind enough to send me a copy of the typescript of this paper, together with a photocopy of Molenaar's paper, referred to in the next note. Collett's paper will be published by FAO in a fully revised edition of Paper no. 60 which is due out shortly.)
52. A. Molenaar, *Water Lifting Devices for Irrigation*, FAO Agricultural Paper no. 60, 1956.
53. Al-Jazarī, *The Book of Knowledge of Ingenious Mechanical Devices*, trans. and annotated by Donald R. Hill (Reidel Dordrecht, 1974), pp. 179-89.
54. Glick, *Islamic and Christian Spain*, p. 238.
55. Silvio A. Bedini and Francis R. Maddison, 'Mechanical Universe, the Astrarium of Giovanni de' Dondi', *Transactions of the American Philosophical Society*, new series, vol. 56,5 (1966).
56. Ahmad Y. Hassan, *Taqī al-Dīn and Arabic Mechanical Engineering* (University of Aleppo Press, 1976 — in Arabic), pp. 57-70.
57. Carra de Vaux, 'Philon de Byzance', p. 210.
58. Hassan, *Taqī al-Dīn*, pp. 39-40.

9 Power from Water and Wind

Water Power

For about 2,000 years water was the main source of power for milling and other industrial purposes. For about half that period, apart from the muscular energy of men and animals, it was the only source. Even today, given the right hydraulic conditions, water power can compete successfully with other sources of energy. In view of the importance of water power in the social and economic life of many communities, it is hardly surprising that a good deal of attention has been paid to the subject by historians of technology. As we shall see, the use of water power began in about the second century BC, and we might therefore expect, from this comparatively recent date, that scholars would have been able to resolve the main problems about the origins of diffusion of the various types of water-powered machines. This is not the case, partly because the evidence is very scanty for certain periods; as usual, there is very little information from the Dark Ages and any conclusions about applications of water power in those centuries are bound to be tentative. We do now have a reasonably clear picture about the development of water-based industries in medieval Europe from the eleventh century onwards, but until now very little has been published on the situation in Islam. Some historians, assuming from the lack of available information that the Muslims made only limited use of water power, have constructed theories to explain this apparent lack of interest as being due to factors inherent in Muslim society. These theories are, however, based upon a faulty premise, since it can be demonstrated that the Muslims were anything but indifferent to the benefits to be obtained from the exploitation of water power.

Water-Wheels

Before the origin and diffusion of water-wheels can be discussed, it is necessary to describe the three basic types of wheel. The vertical, undershot wheel is a paddle wheel which is installed on a vertical axle

155

over a running stream (Figure 9.1). Its power derives almost entirely from the velocity of the water, and it is therefore affected by seasonal changes in the rate of flow of the stream over which it is erected. Furthermore, the water level may fall, leaving the paddles partly or totally out of the water. The efficiency of the undershot wheel is not high — perhaps as low as 22 per cent — because so much of the energy of the water is dissipated by turbulence and drag. The fact that the undershot wheel retained its popularity over many centuries is due to the simplicity of its construction and to special measures that can be taken to increase its performance. These will be discussed later.

The overshot wheel is also vertical on a horizontal axle. Its rim is divided into bucket-like compartments into which the water discharges from above, usually from an artificial channel or 'leat' (see Figure 9.2). Its efficiency can be as high as 66 per cent, provided all the water from the leat falls into the buckets and there is no spillage.

Figure 9.1: Undershot Water-wheel

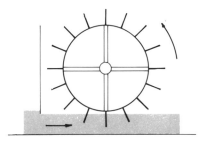

Figure 9.2: Overshot Water-wheel

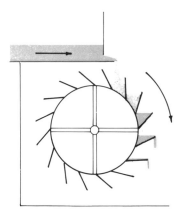

Figure 9.3: Gears for Mills

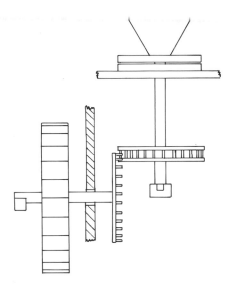

When used for corn milling both types of vertical wheels require a pair of gears to transmit the power to the millstones. A vertical toothed wheel is mounted on the end of the water-wheel's axle inside the mill house. This engages a lantern-pinion, whose vertical axle goes up through the floor to the milling room; it passes through the lower, fixed millstone and is fixed to the upper, rotating stone. The corn is fed into the concavity of the upper stone from a hopper (see Figure 9.3).

The third type of wheel is horizontal, and can be subdivided into two main types. In the first of these a wheel with curved or scooped vanes is mounted at the bottom of the vertical shaft and water from an orifice fitted to the bottom of a water tower is directed on to the vanes, the flow being therefore mainly tangential (see Figure 9.4). The second type is made by cutting along the radii of a metal disc, then bending the segments to form curved vanes, rather like those of a small, modern air-fan. The wheel, also fixed to the lower end of a vertical axle, is installed inside a cylinder, into which the water cascades from above, turning the wheel mainly by axial flow. The efficiency of wheels of the first type — i.e. those driven by tangential flow — has been estimated at about 40 per cent for mills in modern Crete with a power of one kilowatt delivered to the millstone shaft.[1]

Figure 9.4: Horizontal Water-wheel

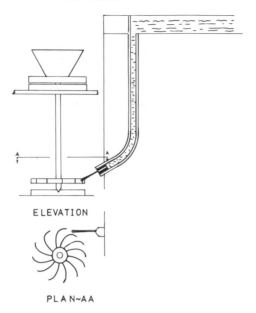

ELEVATION

PLAN~AA

One of the main problems, in any discussion of the history of water-wheels, is that the sources seldom tell us unequivocally which type of wheel they are referring to. The oldest known water-wheel, which was used for grinding corn, was the *hydraletes* described by Strabo about 24 BC as existing at Cabeira in the Pontus, having formed part of the property which the last Mithridates lost when he was overthrown by Pompey in 65 BC.[2] The first literary reference occurs in a Greek epigram attributed to Antipater of Thessalonica, which was probably written about 30 BC. This often-quoted passage celebrated the release of women from the drudgery of using hand querns. It includes the lines:

> Your task is now for the nymphs, by command of Demeter,
> And leaping down on the top of the wheel, they turn it,
> Axle and whirling spokes together revolving and causing
> The heavy and hollow Nisyrian stones to grind above.[3]

This has sometimes been taken as meaning that the writer was referring to a horizontal wheel, but the lines could apply equally well to an overshot wheel.

The first clear description of any type of water-wheel occurs in Vitruvius, whose work on architecture was completed about 27 BC. Significantly perhaps, the passage occurs immediately after a brief description of the noria:

Mill wheels are turned on the same principle, [i.e. as the noria] except that at one end of the axle a toothed drum is fixed. This is placed vertically on its edge and turns with the wheel. Adjoining this larger wheel there is a second toothed wheel placed horizontally by which it is gripped. Thus the teeth of the drum which is on the axle, by driving the teeth of the horizontal drum, cause the grindstone to revolve. In the machine a hopper is suspended and supplies the grain, and by the same revolution the flour is produced.[4]

The machine is therefore a corn mill driven through a pair of gears by an undershot wheel; Vitruvius makes it clear that this type of wheel is intended by his reference back to the paddle-driven noria. The wheel is sometimes called 'Vitruvian' and some writers have called the wheel and the mill which it operates Roman inventions, because they were first described by Vitruvius. Vitruvius, however, is often at pains to point out that the machines that he describes are in everyday use, and he frequently attributes the invention of particular devices to Greek or Hellenistic engineers. For the moment let us leave the undershot wheel and turn to the other two types.

The earliest evidence we have of the overshot wheel is the occurrence of such a wheel on a mural in the Roman catacombs dated to the third century AD. The remains of a fifth-century overshot wheel were discovered in the Athenian Agora in 1936.[5] Otherwise there is very little information before the Middle Ages, when references to the overshot wheel occur in Arabic writings. An eleventh-century Spanish Muslim, for example, gives the correct advice that the overshot wheel is to be preferred to the undershot type if the flow of water is scanty.[6]

Evidence for the early existence of the horizontal wheel is even scantier; indeed, if the available data is rigorously examined, there is no real proof for its existence before the eighth century, when confirmation of a wheel of this type is found in the description of a water-mill in a collection of Irish tracts called *Senchus Mor*.[7] It may have been in use in Ireland a century or so before this time. Writing in Baghdad about AD 850, the Banū Mūsà, in their *Book of Ingenious*

Devices describe an alternating fountain operated by a horizontal wheel. This is an axial flow type, since it is rotated from below by a ring of vertical jets.[8] It is reasonable to assume, when a component is described in a treatise of this nature, that it had been in everyday use for some time, so wheels of this type may have been the prime movers for mills in the Near East in the eighth century or earlier. But except for the reference in the Banū Mūsà and in the treatise of al-Jazarī (written in 1206),[9] the axial-flow wheel was not known in Islam — at least there in no evidence for it. The traditional type of wheel, still in use in Iran and elsewhere in this century, was the tangential-flow type. At Tamworth, England, the remains of a tangential-flow wheel dated to about AD 850 have been discovered, in an area where the hydraulic conditions would have favoured the use of a vertical wheel.[10] Finally, there exists a treatise, translated from Greek into Arabic, by a certain Apollonius, called by the Arab biographers 'the carpenter, the geometrician'.[11] He is said to have been from Alexandria, but it is not known when he lived, although this was probably in pre-Islamic times. His treatise describing a musical automaton, refers to a water-wheel which, he says, may be of the horizontal 'Byzantine' type or the 'newly-invented' vertical wheel.[12] If the treatise could be dated to (say) the third century BC, which is possible but quite unverifiable, then this would imply that the horizontal wheel was known in Asia Minor, and that the vertical wheel came into use at about the same time as the noria.

Turning briefly to the situation in China, as discussed by Joseph Needham, we find that water-wheels were in use no later than the first century AD for operating trip-hammers, and for working the bellows for the smelting of iron.[13] No indication is given in the Chinese sources as to the type of water-wheel used for these purposes, but the traditional type in China was the tangentially-driven horizontal wheel, although the first iconographic evidence for its occurs in a work of AD 1313. One illustration shows a horizontal wheel operating a bellows through a fairly complex linkage for converting rotary to reciprocating motion. Another shows a vertical undershot wheel driving a number of mills through gears, as in the Vitruvian mill.

It is not easy to construct a coherent hypothesis for the origins of the three types of wheel from the scattered references listed above. The weight of evidence suggests that both types of vertical wheel originated in Hellenistic times by analogy with the noria. Vitruvius certainly implies that he realised that the undershot wheel and the noria were closely related and there is also a mechanical similarity between the overshot and water-raising machines. To quote Landels:

> But there is another equally attractive hypothesis — that the overshot wheel was conceived independently of any other type, by simply reversing the action of the bucket-wheel. If one can put power into that machine and get water out of the top, why not put water into the top and get power out of it? In fact, this possibility would be clearly demonstrated each time somebody finished a spell of work treading a bucket-wheel. It would have to be slowly reversed until all its buckets were emptied, and the pull it exerted during that operation would be clearly felt. The bucket-wheel was certainly in use in Vitruvius' time, perhaps for some while before, and though he does not describe an overshot wheel, that might be due to the fact that the undershot type was the only one he had seen.[14]

If we accept a Hellenistic origin for both types of vertical wheel, their subsequent diffusion throughout Europe and the Middle East can be interpreted as a radial transmission from their point of origin, facilitated by the cultural unity of the Roman Empire. Until the third century evidence is very scanty, perhaps because until that time the abundance of slave labour made the extensive use of water power unnecessary. There is considerable evidence for the use of water-mills in Rome itself, after the third century, where water diverted from the Aqua Trajana was used to power a group of mills on the Janiculum.[15] Mills were installed on Hadrian's wall in Britain, on the Moselle, and there was a very large installation early in the fourth century at Barbegal near Arles in southern France. By the sixth century it seems to have been impossible to feed the population of a large city without using water power, as is illustrated by a well-known incident related by Procopius. At Rome in AD 537 the besieging Goths cut the water supply outside the city, thereby depriving the corn mills to such an extent that the people of Rome were threatened with starvation. The Byzantine general Belisarius solved the problem by mounting a series of undershot wheels on barges, the first known reference to floating mills.[16] Further east, there is very little information about the use of mills before the advent of Islam, but we are told that the last Sasānid king, fleeing before the advancing Arabs, was killed at a water-mill on the River Murghab in the year 641.[17] There seems little reason to doubt that milling was a widespread activity in Europe and the Middle East from the third century onwards.

We have no means of telling the relative importance of the various types of wheel from Roman times through the Middle Ages, but it would seem that the undershot wheel was predominant in Europe. Floating mills and mills on the piers of bridges were very common installations, and these must have used undershot wheels. Because of its greater efficiency the overshot wheel became the most widely-used type from the eighteenth century onwards, in situations where the cost of the hydraulic works could be justified by the return on capital. Even so, the undershot wheel was often the best answer in certain conditions. It is cheaper to build and install than the overshot type and in a fast-running stream with a reliable flow its power output can be very high, despite its low efficiency. To achieve a high output the wheel needs to be large, but this is not a serious problem unless the wheel obstructs a navigable river.

The questions posed by the diffusion of the horizontal wheel do not lend themselves to easy answers. It is likely that they were in use in the Eastern Mediterranean by the first century BC, at about the same time as their first recorded use in China. They were known in northern Europe from the seventh century, and were popular thereafter in many parts of Europe and the Middle East. They were in widespread use, for example, in fourteenth-century Pistoia.[18] Modern authors have described existing installations in Iran.[19] Crete[20] and the Balkans,[21] but although they had probably been in use for centuries in those, and other areas, the actual dates for their introduction to given localities remains unknown. It may well be that they were invented separately in different regions — China, Asia Minor and northern Europe — in communities where there was a need for additional power, and the technical repertoire did not include the use of gears. The axial-flow wheel described by the Banū Mūsā is something of a mystery. It is usually safe to assume, when a component is incorporated in an ingenious device, that it was already in use in utilitarian machines. It is possible, therefore, that the axial-flow wheel was used as a power source in Islam, but there is no certain evidence for its practical application before the invention of the so-called 'tub-wheel' in sixteenth-century Europe.[22] It would be good to know more about the origin of horizontal wheels in general, because they are the direct ancestors of modern turbines.

Grist Milling

As far as we know, for about the first 800 years of their history, water-mills were used in Europe and western Asia exclusively for grist

milling, i.e. the grindling of corn and other seeds. From the available
evidence a broad pattern of activity emerges: a large expansion in
grist milling in the Roman Empire from the third century onwards,
continued in Byzantium and probably in Iran; the widespread use of
large-scale milling installations in Islam from the eighth century
onwards; a similar expansion in northern Europe, starting about the
end of the tenth century and accelerating markedly in the twelfth. In
parallel with these large state-sponsored or capitalistic enterprises,
there were always small mills built by communities to serve only their
own needs. And between these two extremes, and overlapping them,
were the mills owned by lay and clerical landlords, to which the
peasants were obliged to take their corn for grinding — and to pay for
the service. It is not always possible to distinguish among the three
categories, and the pattern changed over the centuries, with large
installations run for profit becoming common in Europe in the later
Middle Ages. The mill played an important part in the daily life of
most communities, but the communal aspects of water power are the
province of social and economic historians. Here we are concerned
mainly with the exploitation of a particular type of energy, and the
efforts made to utilise the power of water to the greatest effect.

 Apart from the mills serving Rome and other large cities, milling in
the later Roman Empire seems often to have been associated with
military needs. The mills along Hadrian's wall have already been
mentioned; at least two of these are known from archaeological
discoveries, both of the Vitruvian type. One, possible dating to the
third century is at Chollerford Bridge near Chesters and the other
over the River Irthing at Willowford Bridge.[23] Millstones have been
found on the Roman frontier fortifications in Germany, notably at
the fort of Saalburg in the Taunus range, which has yielded a great
number of millstones also dated to the third century.[24] By far the most
impressive of the Roman milling installations, however, was erected
at Barbegal, near Arles in Provence. Water from the River Durance
was conducted along an aqueduct down a steep hill with a gradient of
30 degrees, with a fall of about 19 metres. The water was divided into
two millraces, each of which operated eight water-wheels. Some of
the results arrived at by Sagui in his paper on this installation must be
treated with caution, particularly with regard to its input and output
power. He makes a number of assumptions, all subject to errors, the
total effect of which could invalidate his calculations. Nevertheless,
his estimate of a total daily production of 28 tonnes of flour in a 24-
hour period, based upon a comparison between the actual produc-

tion of modern millstones of known size with the actual size of the Barbegal millstones, may be of the right order of magnitude. This quantity would have sufficed for the daily needs of about 80,000 people, whereas the nearby city of Arles had a population of perhaps 10,000. It is inferred, therefore, that the surplus was in part exported from Arles, then an important port on the Rhône, and in part went to supply the Roman troops in southern Gaul. It is known that there were other large installations of a similar nature. At Tournus in Burgundy, for example, there was also a port for the cereals of the Saône, which were utilised by the Roman army.[25]

From scattered references we can infer that corn milling continued in the Byzantine and Sasānid Empires after the fall of Rome. The building of ship-mills on the Tiber in AD 537 and the existence of a water-mill in eastern Iran in 641 have already been mentioned. Al-Mas'ūdī tells us that when the Sasānid king Anushirwān (AD 531–79) invaded Asia Minor he built water-mills on a river to supply his army.[26] Although our first information about corn milling in Islam is from the tenth century, the number of mills in use by that time allows us to suppose that the tradition was well established in pre-Islamic times.

The Muslim geographers and travellers leave us in no doubt as to the importance of corn-milling in the Islamic world. This importance is reflected not only in the widespread occurrence of mills, from the Iberian peninsula to Iran, but also in the very positive attitude of the writers to the potential of streams for conversion to power. The Tigris at its source, says al-Muqaddasī, would turn only one mill,[27] and al-Istakhrī, looking at a fast-flowing stream in the Iranian province of Kirman, estimates that it would turn 20 mills.[28] It is as if these travellers were rating streams at so much 'mill-power'. It is possible to mention only a few of the many references to mills in the works of Arabic writers from the tenth century onwards. At Nishapur in Khurasan there were 70 mills on the river near the city; Bukhara was noted for the number of its mills driven by undershot wheels,[29] and there were many corn mills in the Caspian province of Tabaristan. In the Iranian province of Fars the mills were owned by the State, and there were many mills in all the other Iranian provinces. The town of Bilbays, on the eastern branch of the Nile in Lower Egypt, was an important grain-processing centre. The many mills there ground corn for export to the Holy Cities of the Hejaz, and al-Muqaddasī estimated that its annual output of grain and flour amount to 3,000 donkey loads (say 300 tons).[30] The use of water power was widespread in North Africa, notably at Fez and Tlemcen.[31] In tenth-century Palermo in Sicily, then under Muslim rule, the banks of the river below the city were lined with mills.[32] There are many references to mills in the Iberian peninsula, for example at Jaen and at Merida.[33]

The Muslims used various methods for increasing the rate of flow of the water which operated mills, and so increased the power and the productivity. One such method was to install the water-wheels between the piers of bridges to take advantage of the increased rate of flow caused by the partial damming of the river.[34] Dams were also constructed to provide additional power for mills and water-raising machines, such as the dam built by 'Adud al-Dawla over the River Kur in Iran (see Chapter 3). At Cordoba in Spain: 'across the river, below the bridge, was a dam made of Quptiyya stone, with large marble pillars. In the dam were three millhouses and in each house were four mills.'[35] Until quite recently its three millhouses still functioned, but much changed from their original form.[36]

The ship-mill was widely used in the Islamic world as a means of taking advantage of the faster current in midstream, and of avoiding the problems caused to fixed mills by the lowering of the water-level in the dry season. Mills of this type are reported from Murcia and Saragossa in Spain, from Tiflis in Georgia, and from a number of other places. The most impressive of them, however, were in Upper Mesopotamia, which was the granary for Baghdad:

The ship-mills on the Tigris at Mosul have no equal anywhere, because they are in very fast current, moored to the bank by iron chains. Each has four stones and each pair of stones grinds in the day and night 50 donkey-loads. They are made of wood and iron — sometimes of teak. At Balad, 7 farsakhs from Mosul, there were a large number of these working to supply Iraq, but the tyranny of the Hamdanids left none, nor any people. There were a certain number of them on the Tigris at Haditha; the Hamdanids took them over with their revenues — the revenue was about 50,000 dinars. The mills at Qal'at Ja'bar and Raqqa on the Euphrates did not rival these and were less numerous.[37]

The Hamdanids, who provided minor dynasties in Mosul and Aleppo in the tenth century, have in general been regarded favourably by Arab and European historians, but Ibn Hawqal's attitude to them is one of implacable animosity. We do not known the reason for this, in a writer who usually expressed his opinions with moderation. Even so, it is clear from his account that a kind of factory milling was necessary on the Tigris and the Euphrates in order to meet the needs of Baghdad and the other cities of Iraq — and that it was a very lucrative business. If we take a donkey-load as 100 kg, then the output of one of these mills in 24 hours was 10 tonnes, sufficient for about 25,000 people. At this time the population of Baghdad has been estimated at 1.5 million, making this kind of large-scale milling absolutely essential.[38] The population consisted of a large bureaucracy, a standing army, merchants, artisans, scholars, shopkeepers and labourers. It has been suggested that the Muslims were slow to use water power (and we have seen that this is not the case) because the abundance of slaves made mechanisation unnecessary. Slaves were indeed plentiful at certain times, but they were used as domestic servants, concubines and soldiers. Long after the time of Ibn Hawqal, Upper Mesopotamia continued as a large supplier of flour to Iraq. In about 1183 Ibn Jubayr saw the ship-mills across the River Khabur, 'forming, as it were, a dam'.[39]

Further evidence of the Muslims' eagerness to harness every available source of water power is provided by their use of tidal mills. This application is, of course, not possible in the Mediterranean, but in the tenth century in the Basra area there were mills that were operated by the ebb-tide. 'The ebb-tide is also useful for operating the mills, because they are at the mouths of the rivers, and when the water comes out it turns them.'[40]

The quality of millstones has always been important. They must be hard but of homogeneous texture, so that pieces of grit do not become detached and get mixed with the flour. The stone from certain localities was therefore particularly prized for milling purposes. In the area of Majjana, in modern Tunisia, for example: 'They cut millstones from the nearby mountains, which are excellent and good for grinding; they can last a man's lifetime without dressing or other treatment, due to their solidity and the fineness of their grains.'[41] The black stone of al-Jazira — i.e. Upper Mesopotamia — was called 'the stone of the mills'. It was the stone used for the mills that supplied Iraq with flour. It had no equal; a single millstone made from this material cost about 50 dinars.[42] Stone for the mills of Khurasan was mined from the hills near the city of Herat.[43]

It is usual to begin the story of European corn-milling with a mention of the 5,624 mills recorded in the Domesday Book. This is an impressive figure at first sight, and indeed it does provide evidence for a commitment to the use of water power for the processing of grains. It should be borne in mind, however, that the population of England at that time was of the order of one million,[44] so that on average less than 200 people were provided for by each mill. The mills must therefore have been small, low-powered units and it seems very likely that most of them were operated by horizontal water-wheels. There is no reason to doubt that France and Germany were less advanced than England in the use of water power, but for these countries we lack the detailed documentation that is furnished by the Domesday Survey of 1086. There were certainly mills in France by the close of the tenth century, and the subsequent history of large-scale milling in France indicates that the extension of milling during the twelfth century was based upon an established tradition, a tradition that had probably continued at a reduced level since the end of the Roman Empire.

The position in Spain is complicated by the coexistence of the two cultures — Muslim and Christian — over a period of more than seven centuries. The widespread use of mills throughout the Muslim world in the tenth century, including the Iberian peninsula, is well attested. Most of these seem to have been driven by undershot wheels — ship-mills and mills on the piers of bridges must have been of this type. On the other hand, there is ample documentation attesting to the build-ing of mills in the Cantabrian and Pyrenean mountains in the ninth century and all over Christian Spain in the tenth century. Mills were especially features of the Catalan landscape, for which profuse documentation exists from the ninth century onwards.[45] The horizontal mill was known in Spain from as early as AD 800. the vertical mill probably not until the mid-tenth century. Subsequently, there may have been a progressive displacement of horizontal by more powerful vertical mills. Nevertheless, the horizontal mill, because it was economical to build and maintain, must have pre-dominated throughout the Middle Ages.[46]

In Europe the methods used by the Muslims for increasing the power delivered to mills found widespread application. Outside Spain the earliest clear reference to the use of dams is the damming of the River Leck about AD 1000 to supply water mills at Augsburg.[47] Ship-mills were in use in a number of French cities by the same date,[48] and tide-mills are first mentioned in the eleventh century and sub-sequently have a more or less continuous history to modern times.[49] There is no evidence to suggest that any or all of these developments was due to the direct influence of Islamic practices. All the techniques except tide-mills were known in pre-Islamic times, although only for the horizontal and undershot wheels is there evidence for diffusion into northern Europe before the Middle Ages. It seems likely that there was an independent development of water-power technology in Europe, but that this was stimulated and enriched by Muslim ideas, probably transmitted from Muslim Spain.

Two examples, both from France, will have to suffice as illustrations of the power-consciousness of Europeans, always present, but becoming obsessive from the twelfth century onwards. On the River Garonne, near Toulouse in France, there were at least 63 floating mills, but these had certain disadvantages, in that they were an obstacle to navigation and tended to break away from their moorings in times of flood. Three large dams were therefore built across the river, and 43 mills were erected on its right bank. The dams were at Château-Narbonnais (16 mills), La Daurade (15 mills) and Bazacle (12 mills) — in the order upstream to downstream. The height of a dam determines the power; the higher it is the greater will be the head of water and hence the grain will be ground faster. But if a downstream dam is heightened those upstream will suffer, because they will not have a head of water sufficient to drive their water-wheels. From the completion of these installations late in the twelfth century, and for almost another two centuries, there were frequent legal wrangles, as the owners of the lower dams raised the crests of their dams to the detriment of those upstream. Progressively, however, the shareholders came to realise that it was in their interests to share their profits and losses, and in the 1370s the Chateau-Narbonnais and the Bazacle associations had formed themselves into what we should call limited companies, the Daurade association having ceased its activity after a long lawsuit with Le Bazacle. These two medieval companies moved smoothly into modern times. In the nineteenth century the Bazacle dam, having been rebuilt in 1709 after the medieval dam had been destroyed by a flood, was used to produce electricity. After the Second World War, the company was nationalised by the French government.[50]

In Paris, where there were no floating mills, the preferred location for mills was in between or just below the piers of bridges, since it was here that the flow was fastest to deliver the greatest power to the water-wheels. In the thirteenth century, in addition to the ten mills at the Pont-aux-Meuniers between the right bank and the great arch left open for navigation, there were at least 55 in the waters of the Seine between the tip of Île-Notre-Dame and the Pont-aux-Meuniers. The positioning of mills at bridges caused damage by scouring to the piers and was also a severe hindrance to navigation. There were therefore frequent disputes between the mill-owners on the one hand and the boatmen and bridge-owners on the other.[51]

Industrial Uses of Water Power

In AD 751, after the battle of Atlakh, Chinese prisoners of war introduced the industry of paper-making in the city of Samarkand. The paper was made from linen, flax or hemp rags after the Chinese method. Soon afterwards paper-mills on the pattern of those in Samarkand were erected in Baghdad, the Yemen, Egypt, Syria, Iran, North Africa and Spain.[52] It has already been mentioned that the Chinese were using water power for industrial purposes by the first century AD, and there is abundant evidence that water-powered trip-hammers were used in China by the third century.[53] It is therefore probable, but not certain, that the early paper-mills in Islam used trip-hammers operated by vertical undershot water-wheels to pound the raw materials. In this system a number of cams are attached to the extended horizontal axle of the wheel. As the axle rotates, the cams bear down in succession on the pivoted lever-arms of the trip-hammers; when the cam moves away the hammer falls on to the material. In a treatise on minerology, written between 1041 and 1049, al-Bīrūnī gives the following description for the processing of gold ores:

The gold may be combined with stone [i.e. ore] as if it were cast with it, so that it needs pounding. And mills pulverise it, but pounding it by *mashājin* is more correct and a more refined treatment — it is even said that this increases its redness, which if true is strange and surprising. The *mashājin* are stones which are fixed to axles that are erected across running water for pounding, as is the case in Samarqand with the pounding of flax for paper.[54]

The description is too brief to enable us to say with certainty that the devices referred to were water-powered trip-hammers, but this is indicated by the reference to a horizontal axle. It would be surprising if the Chinese who built the paper-mills in Samarkand in AD 751 did not make use of this technology, which has been known in China for several centuries. It is generally assumed, but not proved, that the mills at the paper-making centre at Jativa in Muslim Spain were driven by water.[55]

There are other references that attest to the industrial use of water power in Islam. Writing in 1107, Ibn al-Balkhī calls a newly-restored dam on the River Kur in Iran by the name of Band-i-Qassār, meaning 'Fullers' Dam', an indication that its impounded water provided power for fulling mills.[56] In a recent survey in the Jordan valley, the remains of 32 water-powered sugar-mills, dating to the Ayyubid-Mamlūk period, were discovered.[57] Writing in the first half of the twelfth century, the historian Ibn al-'Asākir mentions the use of water power for sawing timber. In some of the devices, such as water-clocks, described by al-Jazarī in his machine book completed in 1206, there are small water-wheels with cams on their axle which are used to operate automata. These were probably copied from the similar mechanisms in industrial mills, which were used to activate trip-hammers.[58] A great deal of research, both literary and archaeological, needs to be done before an assessment of the extent of industrial milling in Islam can be made. At present the information is sparse, and scattered among a wide variety of sources. Until about the middle of the twelfth century, the same can be said of Europe, and some of the evidence for early beginnings of industrial milling is rather tenuous.

Industrial milling probably began in northern Europe at the end of the tenth century. Lynn White introduces the topic as follows:

And in fact in the very late tenth or eleventh century we begin to get evidence that water-power was being used for processes other than the grinding of grain. By 983 there may have been a fulling mill — the first useful application of the cam in the Occident — on the banks of the Serchio in Tuscany. In 1008 a donation of properties to a monastery in Milan mentions not only mills for grinding grain, but, adjacent to them along the streams, *fullae* which were probably fulling mills. In 1010 the place-name Schmidmulen in the Oberpfalz indicates that water-driven trip-hammers were at work in Germany. About 1040 to 1050 at Grenoble there was a fulling mill, and *c.* 1085 one for treating hemp. By 1080 the Abbey of Saint Wandrille near Rouen was receiving the tithes of a fulling mill, and by 1086 two mills in England were paying rent with blooms of iron, indicating that water power was used at forges. Before the end of the eleventh century, iron mills were likewise found near Bayonne in Gascony.[59]

Comparing this data with the mentions given above for Islam, it can be seen that in neither case do we have firm evidence for the introduction of industrial milling before the middle of the eleventh century, but in both societies it would be reasonable to assume that a start had been made before that time.

There is sufficient evidence for a rapid growth of industrial milling in Europe from the twelfth century onward, and indeed it would not be possible even to list all the activities that benefited from the application of water power during the Middle Ages.[60] Among the most important were the fulling of cloth, paper manufacture, iron forging, sawmills[61] and tanning. The large monasteries were particularly interested in the building of water mills, especially the Cistercians. Their abbey in southern Champagne, to take but one example, had water-driven fulling, paper- and iron- mills in the thirteenth century. To quote Forbes:

They were particularly involved in the building of water-driven hammer forges and they appear as pioneers in iron metallurgy with the help of water-mills not only in France but also in Germany, Denmark and Great Britain. Out of 30 French documents of the twelfth century concerned with hammer forges and iron metallurgy 25 were drawn up by Cistercian monks and producers. Some of these abbeys were water-driven workshops combining very different crafts under one roof.[62]

Industrial mills appeared in Christian Spain, notably in Catalonia, during the twelfth century, the period of expansion of milling technology throughout western Europe. There are frequent citations of fulling mills in Catalonia from 1150 on, and towards the end of the century water power was applied to the Catalan forges. Paper-mills also appear in Catalan documentation during the 1150s and, although there is no hard evidence that the mills themselves were of Islamic origin, there is no reason to doubt they were not, inasmuch as the rest of the technology of paper-making was identical with Muslim techniques. Glick suggests that the prototype of the trip-hammer fulling mill was the Chinese rice-husking mill, which was vertical and undershot, and that it might have diffused to Muslim Spain along with the introduction of rice. Its use could have been applied to other industries there and have diffused northward.[63] There is no reason why the processing of rice should have been the first industry in al-Andalus to use water technology; the priority could equally well have been with the paper or sugar industries, although this does not affect Glick's main argument.

The question of the diffusion of industrial milling cannot be easily answered. If the technology had passed from Muslim Spain to Catalonia and from there into northern Europe, we should expect that industrial milling would have begun earlier in Catalonia than in other parts of Europe, but this does not appear to have been the case. We have no positive proof that industrial milling began anywhere in Islam before the eleventh century. The balance of probability, however, is in favour of the technology having been developed in Islam after the period of expansion and consolidation had been completed about the middle of the eighth century, and of a fairly slow diffusion of industrial milling throughout the Muslim world during the ensuing centuries. If this were so, there is no reason why diffusion to western Europe should not have been through the Byzantine lands, rather than by the route from southern to northern Spain, and thence into France. This is, of course, conjectural, and the possibility of separate developments in Islam and Europe cannot be ruled out.

Windmills

The windmill as a prime mover was probably unknown to the Greeks and Romans. Hero of Alexandria (*fl.* mid-first century AD) describes a wind-organ, the piston providing the air for the organ pipes being moved by a wheel 'which has oar-like scoops as in the so-called wind-motors'. The Greek word is *anemourion* (wind-vane), a term which, except in this text, occurs only in the writings of a twelfth-century bishop. Some of the drawings in the Hero manuscripts show a kind of windmill moving the piston, but they are later additions, probably made by Christian or Muslim copyists who knew the windmill proper.[64] No other classical writer mentions windmills, and Hero's little wind-wheel was probably a product of his inventive mind. The Banū Mūsà, writing in Baghdad about AD 850, also mention a small wind-wheel, in this case used to operate an alternating fountain. They certainly knew Hero's works, and the Arabic word used for the wheel is clearly a corruption of *anemourion*.[65] It is safe to assume that their wheel was derived from Hero's, a derivation which implies strongly that the anemourion was in the original Greek text.

The first reference we have to full size windmills occurs in the writings of al-Istakhrī, who mentions the windmills of Seistan (the western part of modern Afghanistan).[66] Al-Istakhrī's book was completed before 951, but in one of al-Mas'ūdī's books, written a few years later, he related a story of a Persian who claimed to the Caliph 'Umar I (634-44) that he was able to build a windmill. 'Umar made him substantiate his claim by building one.[67] The story is somewhat unreliable, because there are tendencies in some of the historians of the ninth and tenth centuries to invent traditions that show the Persians in a better light than the Arabs. But while we must accept al-Istakhrī as our first reliable witness, he was probably describing a tradition which had existed for some time before his report. The earliest description of these windmills is given by the Syrian geographer al-Dimashqī, in a book written about 1271. He tells us that the mills were supported on substructures built for the purpose, or on the towers of castles or on the top of hills. They consisted of an upper chamber in which the millstones were housed and a lower one for the rotor. The axle was vertical and it carried twelve or six arms covered with a double skin of fabric. The walls of the lower chamber were pierced with funnel-shaped ducts, with the narrower end towards the interior in order to increase the speed of the wind when it flowed on to the sails.[68] This type of windmill spread throughout Islam, and to China and India. In medieval Egypt it was used in the sugar-cane industry, but its main application was to grist milling.

The earliest authentic record of a windmill in western Europe yet discovered is in a deed, reliably dated as *c.* 1180, recording the gift of land near a windmill to the Abbey of St Sauvere de Vicomte in Normandy.[69] In 1185, at Weedley in Yorkshire a windmill was rented for eight shillings a year. Before Henry II's death in 1189, one of his constables gave a windmill near Buckingham to Oseney Abbey, and in 1191 or 1192 Jocelin of Brakelond mentions one as though it were no novelty. Its complete integration with medieval society occurred when Pope Celestine III (1191-8) ruled that windmills should pay tithes. During the next 100 years windmills became one of the most typical features of the great plains of northern Europe. During the thirteenth century, for example, 120 windmills were built in the vicinity of Ypres alone.[70] They had certain advantages over water mills: in the winter their operation could not be stopped by freezing and they could be used to grind grain for besieged castles. In Syria, the great Crusader castle of Krak des Chevaliers, completed in 1240 and still largely intact, had a windmill on its walls.[71]

The first known illustration of a windmill occurs in an initial letter of the 'Windmill Psalter' written at Canterbury about 1270 and now in

New York. This shows a post-mill of orthodox construction, which was, it may safely be assumed, a feature of the landscape familiar to the artist who painted it.[72] By the second decade of the thirteenth century the windmill was familiar enough in Italy for Dante to use it as a metaphor in describing Satan threshing his arms 'come un molino che il vento gira'.[73] It was, however, quite a late arrival in Spain, where it seems not to have been introduced until the sixteenth century.[74]

Medieval windmills were post-mills, in which the whole structure is rotated in order to face the sails into the wind. The tower mill, having a rotatable cap with a fixed substructure, may have been introduced towards the end of the fourteenth century but did not come into general use for about another two centuries. Working post-mills, not materially different from those shown in medieval illustrations, are still to be found in Britain and elsewhere in Europe. They

Figure 9.5: Windmill

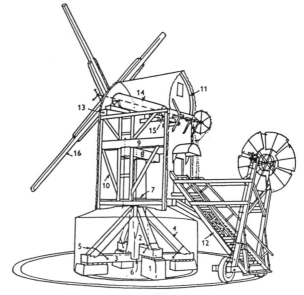

A typical East Anglian post-mill

1. Brick pier	7. Centering wheels	12. Steps or ladder
2. Main-post	8. Crown-tree	13. Weather-beam
3. Cross-tree	9. Side-girt	14. Wind-shaft
4. Quarter-bar	10. Brace	15. Tail-beam
5. Retaining strap	11. Cap-rib	16. Sail-stock
6. Heel of main-post		

are, of course, no longer economically viable, but are kept in order as living evidence of our industrial past.

By any standards, the post-mill must rank as one of the most daring inventions in the history of engineering. The concept of balancing a complete engine-house, together with its prime mover, on a single fulcrum would, if proposed today, be dismissed as impracticable or downright dangerous. Yet machines of this type were in service as an important source of energy for over 700 years. The following passage is a condensed version of the description of a typical East Anglian post-mill by Stanley Freese (see Figure 9.5):

> Post-mills are mounted upon four (or sometimes six) piers of stone or brick (1). One pair of opposite piers is normally higher than the other pair, so that the timber cross-trees (3) can pass over one another at the centre. The mill revolves about the main-post (2). The main-post, usually a great baulk of oak, 2 feet 6 inches square, 18 feet high and weighing about $1\frac{1}{2}$ tons, stands above the cross-trees but is *not* resting on them, as is commonly supposed. Nor are the cross-trees secured to the brick piers in any way: they rest upon the piers under the great weight of the windmill, which prevents them from shifting during a storm. They ought to be of oak, about 20 feet long, 12 or 14 inches deep, and 10 or 12 inches thick, and should have about an inch of clearance betwee them; this clearance is for circulation of the air, and to allow for settling ...
>
> Holding up the main post are the four oak quarter-bars (4), 10 or 11 inches square and weighing about 3 cwt. each. These are tenoned and housed into the main-post at their heads, and into the cross-trees at their foot ...
>
> The top of the main-post is the chief bearing, about which the mill turns; it terminates in a pintle, onion-head or cock-head, turned out of the solid wood of the post.[75]

Freese continues to describe this type of mill in great detail. The crown-tree is a great baulk of timber which transfers the load of the mill to the main-post — the pintle of the main-post enters a socket cut in the centre of the crown-tree. Each end of the crown-tree supports the side-girts or sheets (9), which weigh about half a ton each and measure 12–15 inches deep and 9 inches thick. The rest of the framework, floors and cladding of the mill is built around these timbers. Typically, post-mills have a floor space of 16 or 18 feet by 10 feet, and measure 22 to 25 feet from the bottom floor to the roof ridge. Access is by means of a ladder (12), which is provided with wheels so that it moves easily when the mill is luffed into the wind. The sails, usually four, are carried on a massive wooden axle, at the end of which is a vertical gear that meshes with a horizontal gear upon whose axle the millstones are installed. Usually a secondary set of gears operate the sack hoist.

There has been considerable discussion about the origins of the European windmill, in particular as to whether it was diffused from Islam or was a completely independent invention. Its late arrival in Spain indicates that it was transferred there from northern Europe, and mills with vertical sails were not used in medieval Islam. It seems likely that the post-mill was developed in northern Europe by derivation from the Vitruvius water-mill, to which it is closely related mechanically. The *idea* of using wind as a source of power may, however, have come to Europe from Islam.

Historians have often expressed surprise about the absence of the vertical-sailed windmill from the Muslim world, since it is a much more powerful machine than the horizontal-sailed type. There are good reasons for this, not the least being that water is a much better source of energy than wind. Although there are a few cases of windmills used for industrial purposes, they are best suited to the operations of grist milling and land drainage, in which the fluctuations in the power supply do not pose serious problems. Even for these two purposes, however, water was usually preferred when it was available. In a recent survey of mills in the county of Hampshire — mostly remains but a few still working — 192 water-mills and 8 windmills were counted.[76] Hampshire is an agricultural county, well supplied with chalk streams, and it would seem that windmills were only built where there was no water nearby. Several factors must have influenced the decision to build a windmill: one of these, obviously, was the reliability of the winds, another the local hydraulic conditions, and a third the availability of suitable timber. Non-technical considerations must also have been weighed. If transport was expensive, there would have been a critical distance for the return journey to a water-mill; beyond that distance it would have been cheaper to install a windmill. Also, a landowner whose estate did not have any perennial streams might have preferred to build a windmill, rather than pay the owner of the nearest water-mill.

In Islam the great centres of population were (and are) based in river valleys and depend upon irrigation for their agriculture. Obviously, these centres had running water, and as we have seen, water-mills were a common feature of Muslim society. Transport, either by water or by pack-train was fairly cheap, and there seems to have been no difficulty in moving large quantities of flour over long distance, for example from Egypt to the Hejaz or from Upper Mesopotamia to Baghdad. Certainly, there were some areas in the Muslim world which could have benefited from the use of post-mills — the Syrian coastlands and parts of North Africa in particular. The fact that these mills were not introduced to those regions cannot be easily explained, but the shortage of suitable heavy timber may have been the most important inhibiting factor.

Notes

1. N.G. Calvert, 'On Water Mills in Central Crete', *Transactions of the Newcomen Society*, vol. 45 (1972-3), pp. 217-22.

2. Joseph Needham, *Science and Civilisation in China* (5 vols. to date, Cambridge University Press, 1954 onwards), vol. 4, pt. 2 (1965), p. 366.

3. Ibid., p. 366.

4. Vitruvius, *De Architectura* (2 vols., Loeb Classics, London, 1934, this reprint 1970), vol. 2, bk 10, Ch. 5, pp. 305-6.

5. Norman Smith, *Man and Water* (Peter Davies, London, 1976), p. 141.

6. Donald R. Hill, 'A Treatise on Machines' in *Journal for the History of Arabic Science*, vol. 1, no. 1 (1977), pp. 33-44.

7. Smith, *Man and Water*, p. 142.

8. The Banū Mūsà, *The Book of Ingenious Devices*, trans. and annotated by Donald R. Hill (Reidel, Dordrecht, 1979), pp. 226-7.

9. Al-Jazarī, *The Book of Knowledge of Ingenious Mechanical Devices*. trans. and annotated by Donald R Hill (Reidel, Dordrecht, 1974), p. 186.

10. I visited the Castle Museum, Tamworth to inspect the remains of this wheel one blade only, probably fashioned by an adze.

11. E. Wiedemann, *Aufsätze zur arabischen Wissenschaftsgeschichte* (2 vols. Olms, Hildesheim, 1970), vol. 1, pp. 86-7.

12. Ibid., vol. 2, pp. 50-6.

13. Needham, *Science and Civilisation*, vol. 4, pt. 2, pp. 366-408.

14. J.G. Landels, *Engineering in the Ancient World* (Chatto and Windus, London, 1978), pp. 20-2.

15. Smith, *Man and Water*, p. 140.

16. Ibid., pp. 140-1.

17. Al-Balādhurī, *Futūh al-Buldān*, ed. M.J. de Goeje (Brill, Leiden 1866), p. 314.

18. John Muendel, 'Horizontal water-wheels in medieval Pistoia', *Technology and Culture*, vol. 15 (1974), pp. 194-225.

19. Hans E. Wulff, *The Traditional Crafts of Persia* (MIT Press Cambridge, Mass., 1966 — 2nd Printing 1976). pp. 280-3.

20. Calvert, 'on Water Mills'.

21. Louis C. Hunter, 'The Living Past in the Appalachians of Europe', *Technology and Culture*, vol. 8 (1967), pp. 445-66.

22. Smith, *Man and Water*, pp. 165-8.

23. L.A. Moritz, *Grain Mills and Flour in Classical Antiquity* (Oxford

University Press, Oxford, 1958), pp. 136-7.
   24. Ibid., pp. 123-4.
   25. C.L. Sagui, 'La meunerie de Barbegal (France) et les roues hydrauliqes chez les anciens et au moyen age', *ISIS*, vol. 38 (1948), pp. 225-31.
   26. Al-Mas'ūdi, *Kitāb al-tanbīh wa'l-ishrāf*, ed. M.J. de Geoje (vol. 8 of BGA, Brill, Leiden, 1894), p. 139.
   27. Al-Muqaddasī, *Ahsān al-taqālim fī ma'rifat al-aqālīm*, ed. M.J. de Goeje (vol. 3 of BGA, Brill, Leiden, 1906), pp. 136ff.
   28. Al-Istakhrī, *Kitāb al-masālik wa'l-mamālik*, ed. M.G. 'Abd al-'Āl al-Hīnī (Cairo, 1961), p. 166.
   29. Al-Muqaddasī, *Ahsān al-taqālim*, p. 280.
   30. Ibid., p. 195.
   31. Al-Idrīsī, *Description de l'Afrique et de l'Espagne*, Arabic text with French trans. by R. Dozy and M.J. de Goeje (Brill, Leiden, 1866), pp. 75 Ar./86 Fr. for Fez, 80/92 for Tlemcen.
   32. Ibn Hawqal, *Kitāb Sūrat al-Ard*, ed. J.H. Kramers (2nd edn of vol. 3 of BGA, Brill, Leiden, 1938), p. 222.
   33. For Jaen see al-Muqaddasī, *Ahsān al-taqālim*, p. 234; for Merida see al-Idrīsī, *Description de l'Afrique*, p. 222 Ar./183 FR.
   34. Al-Muqaddasī, *Ahsān al-taqaīlim, p. 312.*
   35. *Al-Idrīsī, Description de l'Afrique*, pp. 212 Ar./262-3 Fr.
   36. Smith, *Man and Water*, p. 143.
   37. Ibn Hawqal, *Kitāb Sūrat*, p. 219.
   38. A.A. Duri 'Baghdad' in EI, vol. 1, p. 899.
   39. Ibn Jubayr, *Rihla*, ed. M.J. de Goeje (2nd edn, Brill, Leiden 1907), p. 243.
   40. Al-Muqaddasī, *Description de l'Afrique*, pp. 124-5.
   41. Al-Idrīsī, *Ahsān al-taqālim*, pp. 118 Ar./138 Fr.
   42. Ibn Hawqal, *Kitāb Sūrat*, p. 222.
   43. Al-Istakhrī, *Kitāb al-Masālik*, p. 150
   44. Morris Bishop, *The Penguin Book of the Middle Ages*. (Penguin Books, London, 1971), p. 352.
   45. Thomas F. Glick, *Islamic and Christian Spain in the Early Middle Ages* (Princeton University Press, Princeton, 1979), p. 93.
   46. Ibid., p. 231.
   47. Smith, *Man and Water*, p. 144.
   48. Marjorie Nice Boyer, 'Water-Mills: A problem for the Bridges and Boats of Medieval France' in *History of Technology*, A. Rupert Hall and Norman Smith (eds.) (Mansell, London, 1982), vol. 7, pp. 1-22, p. 7.
   49. Smith, *Man and Water*, pp. 145-6.
   50. Jean Gimpel, *The Medieval Machine* (Gollancz, London 1977), pp. 17-23.
   51. Boyer, 'Water Mills', p. 5.
   52. Cl. Huart and A. Grohmann, 'Kāghad', in EI, vol. 4, pp. 419-20.
   53. Needham, *Science and Civilisation*, vol. 4, pt. 2, pp. 390ff.
   54. Al-Bīrūnī, *Kitāb al-jamāhir fi ma'rifat al-jawāhir*, ed. F. Krenkow (Hyderabad, Deccan, 1936), pp. 233-4.
   55. Glick, *Islamic and Christian Spain*, p. 242.
   56. Norman Smith, *A History of Dams* (Peter Davies, London, 1971), p. 85.
   57. Saleh Hamarneh, 'Sugar-cane Plantation and Industry under the Arab Muslims during the Middle Ages' in *Proceedings of the First International Symposium for the History of Arabic Science* (Aleppo University, 1976), p. 221.
   58. Al-Jazarī, *The Book of Knowledge*, p. 34, p. 105.
   59. Lynn White Jr, *Medieval Society and Social Change* (Oxford Paperbacks, Oxford, 1964), pp. 83-4.
   60. R.J. Forbes 'Power' in Charles Singer, E.J. Holmyard, A.R. Hall and

Trevor L. Williams, *A History of Technology* (5 vols., first published 1956, this reprint 1979), vol. 2, pp. 589-622, pp. 609-10.

61. *The Album of Villard de Honnecourt*, ed. H.R. Hahnloser (Schroll, Vienna, 1935), Table 44.

62. R.J. Forbes, *Studies in Ancient Technology* (5 vols., Brill Leiden, 1955-8), vol. 2 (reprinted 1965), pp. 109-10.

63. Glick, *Islamic and Christian Spain*, p. 233.

64. Forbes in *A History of Technology*, eds. Singer *et al.*, pp. 614-5.

65. The Banū Mūsà (Hill), pp. 222-3.

66. Al-Istakhri, *Kitāb al-masālik*, p. 140

67. Forbes in *A History of Technology*, Singer *et al.* (eds.), vol. 2, p. 615.

68. Al-Dimashqī, *Cosmographie de Chems-ed-Din Abou Abdallah Mohammed ed-Dimichqui*, ed. A.F. Mehren (Arabic text, St Petersburg, 1866), p. 182.

69. Rex Wailes, 'A Note on Windmills' in *A History of Technology*, Singer *et al.* (eds.), vol. 2, pp. 623-8.

70. White, *Medieval Technology*, pp. 87-8.

71. N. Eliséeff, 'Hisn al-Akrād', in El, vol. 3, pp. 503-6.

72. Wailes, *A Note on Windmills*, p. 623.

73. *Inferno*, 24, 1.6.

74. White, *Medieval Technology*, p. 88.

75. Stanley Freese, *Windmills and Millwrighting* (David & Charles, Newton Abbot, England, 1957, reprint. 1974), pp. 17-23.

76. Monica Ellis (ed.), *Water and Wind Mills in Hampshire and the Isle of Wight* (Southampton University Industrial Archaelology Group, 1978), pp. 81-2.

PART THREE:
FINE TECHNOLOGY

10 Instruments

The final part of this book deals with Fine Technology — with instruments, automata and clocks, together with the associated techniques and components. It is convenient to deal with these subjects in separate chapters, although they are closely interrelated in several ways. An urge towards mechanistic explanation led to the construction of simulations of celestial and biological phenomena, and from these much of our technology has evolved, particularly the part embracing fine mechanism and scientific instrumentation.[1] We also frequently find that outstanding engineers designed and constructed all the different devices that are grouped together here under the designation 'Fine Technology' and indeed that the large devices, such as water-clocks, incorporated astronomical representations, biological simulacra and a wide variety of mechanisms and transmission systems. The following three chapters, therefore, proceed from the parts to the whole, but each part, aside from its role in the development of wider systems, also had a distinct purpose and function of its own.

In classical and medieval times astronomy (often associated with astrology) was the queen of the sciences, and the best scholars devoted much of their time to improving observational and computational techniques. Their knowledge — both detailed and profound — is reflected in the sophistication of some of the instruments that they made, and a matching knowledge of early astronomy is required from modern scholars if they wish to understand the purpose and operation of the more complex instruments. It is beyond the scope of this book to give even a summary of the fully developed medieval system, but it is necessary to outline briefly the basic concepts.

In the modern system, the planets, including the earth, follow elliptical paths around the sun, while the moon follows a similar path around the earth. The earth makes one rotation a day on its axis, which is inclined at an angle of about $23\frac{1}{2}$ degrees to the celestial equator. These two principal movements of the earth (there are a number of minor ones that need not concern us here) cause the apparent daily passage of the sun from east to west, due to the earth's rotation, and the apparent movement of the sun, throughout the year, from west to east, along a path called the *ecliptic*, which is inclined at an angle of $23\frac{1}{2}$ degrees to the celestial equator. The band of stars through which the ecliptic passes is called the Zodiac, and it is divided into twelve constellations or 'signs' each occupying 30 degrees. The moon also passes through the Zodiac; its phases are caused by its rotation around the earth. New moon occurs when it is between the earth and the sun, full moon when it is on the far side of the earth from the sun, and the intermediate phases are due to its rotation around the earth. The *synodical* month is the interval between two successive new moons, its average value being 29.53 mean solar days. The sidereal month is defined to be the interval given by the moon's complete circuit of the stars as seen from the earth; its average value is 27.32 mean solar days. If a full moon appears on a certain date, there will be a full moon almost exactly 19 years later — this is the *Metonic cycle.*

The points at which the sun crosses the celestial equator are called the equinoxes, namely the spring or vernal equinox and the autumnal equinox. The longitudes of the sun and the planets at any given time are defined as their angular distance from the vernal equinox, along the ecliptic. Of more general use, however, for problems in spherical astronomy, are the two co-ordinates given to heavenly bodies by reference to the celestial equator. These are the *declination*, the angle between the object and the celestial equator, and the *right ascension*, measured (in hours) eastwards from the vernal equinox along the celestial equator. These values are constant for the fixed stars but vary for the sun and the planets. Because of the sun's apparent daily shift along the ecliptic, the sidereal day is shorter by about four minutes than the solar day. Until the late Middle Ages, it was usual to divide the day (and the night) into temporal or unequal hours; the hours of daylight and darkness were divided by twelve to give 'hours' that varied in length from day to day throughout the year, and from daytime to night on a given day — except at the equinoxes. This system had to be taken into account by the makers of intruments and timepieces.

In our period, the earth was considered to be the centre of the universe with all the heavenly bodies describing circular paths around it. For many problems in spherical astronomy, for example those related to time or to terrestrial co-ordinates, this *geocentric* theory works just as well as the modern *heliocentric* theory. In fact, when carrying out astro-survey work it is simpler to think of one's station as static and the sun or the stars as moving. But problems arise when the movements of the planets are considered, since these cannot be explained by simple circular motion around the earth. Intellectual curiosity was one of the reasons which stimulated men to seek for explanations for these wayward motions. The demands of astrology, however, also meant that accurate forecasts of eclipses and planetary positions had to be possible. The very complex and ingenious system devised to 'save the phenomena' bears the name of Ptolemy. Although it was not invented by the great astronomer of the second century AD, he made the system into a coherent whole, and introduced certain modifications to eliminate inconsistencies. His work was translated into Arabic and became the main foundation of Muslim astronomy, and of medieval European astronomy after its translation from Arabic to Latin early in the thirteenth century.

The bare essentials of the system are shown in Figures 10.1a and 10.1b. In both cases the outer circle represents the Zodiac with the earth (T) at its centre; the smaller 'deferent' circle with its centre at C has an eccentricity of CT with regard to T. The line TC defines the direction of the most distant part of the deferent circle from T, known as the *aux*, and line TCA is referred to as 'line of aux'; V is the vernal equinox. In Figure 10.1a, the sun (S) moves with uniform angular velocity about C, and the angle VTS is the sun's longitude at a given time. Figure 10.1b is for the planets Venus, Mars, Jupiter and Saturn. Ptolemy introduced the equant (E), such that CT = EC, and the centre (O) of the epicycle carrying the planet (P) moves at uniform angular velocity about E. The solutions for Mercury and for

Figure 10.1: Ptolemaic System

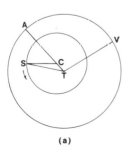

(a)

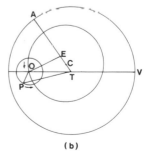

(b)

the moon employ similar principles, but are somewhat more complex. In general, medieval astronomers were interested in computing only the longitudes of the planets, not their latitudes.[2]

The mathematics needed for solving problems in spherical astronomy, even for the simple geocentric system without deferent circles and epicycles, is a laborious business, and it is difficult for the modern mind to comprehend how a scientist such as Ptolemy could have constructed and analysed such an elaborate system without the benefit of place-number numerals, decimals or a fully-developed trigonometry (not to mention logarithms and other modern techniques). Place-number arithmetic and decimals entered the Muslim world from India in the ninth century and came into general use in Islam in the tenth.[3] The Muslims, again basing their work on Indian ideas, also developed trigonometry — plane and spherical — into something approaching its modern form.[4] Even so, the difficulties of computation remained formidable, and astronomers clearly needed some form of assistance if their time was not to be largely consumed in making lengthy calculations. From the time of Ptolemy onwards, one form of assistance was the provision of tables of various kinds, some giving end-results such as the *qibla* tables mentioned in Chapter 7, others listing the values of parameters for use in calculations, for examples tables of trigonometrical and astronomical constants. A large number of such tables (known as *zījes* in Arabic) were compiled by Muslim astronomers. Some of these were transmitted to the West from the twelfth century onwards, but thousands of manuscripts in the libraries of the Middle East have never been published. Significantly, for our purposes, some zijes contain tables for marking out astrolabes and other instruments.[5] In other words, instructions were provided for the construction of the analogue computers which were a necessary aid for practical astronomers (and astrologers).

Calendars

One of the most complex and sophisticated astronomical instruments made before modern times is also one of the earliest known. The Antikythera Mechanism, so called because of its discovery as part of the cargo of a sunken vessel in 1900 near the island of Antikythera (between Crete and Kythera), was probably made in Rhodes about 87 BC. The find consisted of a number of pieces of bronze, much corroded, but clearly showing gear-wheels and parts of inscriptions.

Professor Derek de Solla Price, of Yale University, after 20 years of research, has been able to produce a convincing reconstruction of the mechanism, which is identified as a geared calendar of astonishing complexity.[6] The general plan of all the gearing is shown in Figure 10.2. The gears, all of bronze with teeth cut to equilateral triangles, were mounted on either side of a main bronze plate. It has not yet been possible to identify all the functions of the mechanism, but it is very probable that the contrate wheel A turned the main wheel, and that one rotation of the latter represented a solar year, this being the main input into the instrument. A very interesting part of the mechanism is a differential turntable, the purpose of which may have been to produce as output a function of the Metonic cycle. Other outputs were the positions of the sun and the moon, and the risings

Figure 10.2: Antikythera Mechanism

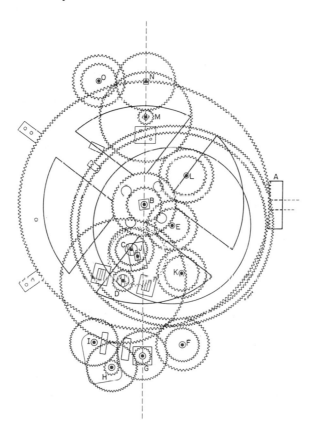

and settings of the circuit of notable fixed stars through the cycle of years and months. The Antikythera Mechanism, to quote Price:

> ... requires us to completely rethink our attitudes toward ancient Greek technology. Men who could build this could have built almost any mechanical device they wanted to. The Greeks cannot now be regarded as great brains who disdained manual labor or rejected technology because of their slave society. The technology was there, and it has just not survived like the great marble buildings, statuary, and the constantly recopied literary works of high culture.[7]

It is worthy of notice that this mechanical technology, not only in the construction of instruments, but also in the fields of water-raising, horology and automata, was not a product of classical Greece, but of the Hellenistic culture with its intellectual centre in Alexandria. As we see in the two final chapters, Arabic scholars, by translating the works of writers such as Ptolemy, Hero and Philo, were instrumental in the transmission of this tradition to Europe. And from the foundation of the Hellenistic writings, they produced their own tradition of fine mechanical technology.

This transmission is exemplified by a geared calendar attributed to the astronomer Abu Saʿīd al-Sizjī. It was described by al-Biruni in a treatise written in the second quarter of the eleventh century.[8] This is not so complex as the Antikythera instrument, although Muslim skills in astronomy and metalwork would certainly have been equal to the construction of such a calendar. As usual, however, we can only record what has survived. Al-Bīrūnī's calendar consists of eight gear-wheels installed on a circular brass plate (Figure 10.3a). Wheels 1 and 2, 3 and 7, 4 and 8 are each single units, soldered together concentrically, as shown in Figure 10.3b. Figure 10.3c shows wheel 7, which is the volvelle for the moon's phases; it has two silvered circles, and its rim, inside the teeth, is divided into 59 divisions. The main plate has a rim, over which a circular lid is fitted. The axles from wheels 1, 5 and 6 pass through holes in this lid. An alidade on the axle of wheel 1 moves round a scale divided into the days of the week. Pointers on the axles of wheel 5 (moon) and wheel 6 (sun) move around scales divided into the twelve signs of the Zodiac, each sign being subdivided into 30 degrees (not shown on Figure 10.3d). The hole for the moon's phases is the same diameter as the silvered circles on wheel 7, so that as the latter rotates the moon's

Figure 10.3: Al-Bīrūnī's Calendar

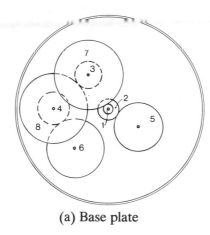

(a) Base plate

Teeth on the gears have been omitted. Figure 10.3 (b) is an 'exploded' section to show the meshing of the gear-wheels. Hole *a* on Figure 10.3 (b) and 10.3 (d) is the same diameter as one of silvered circles on wheel 7.

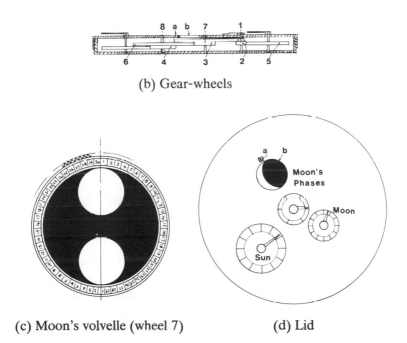

(b) Gear-wheels

(c) Moon's volvelle (wheel 7)   (d) Lid

shape changes daily; one rotation of the wheel corresponds to two lunations. The small hole to the left of the centre-line of the large hole shows the day of the lunation marked on the rim of wheel 7.

The number of teeth on each wheel is as follows:

Wheel	Teeth
1	7
2	10
3	19
4	24
5	40
6	48
7	59
8	59

These numbers give a synodical month of $29\frac{1}{2}$ days and a sidereal month of 28 days. During a complete rotation of wheel 6, wheel 1 rotates 52.35 times, equivalent to 366.42 days. Over a period these slightly inaccurate figures would have produced distortions in the calendar, and we must assume that it was reset from time to time, probably by moving the pointers relative to their axles. Al-Bīrūnī says that the wheels on his calendar are like those fitted to the backs of astrolabes. Geared astrolabes have survived in a very few cases, but the survivors are later than the time of al-Bīrūnī.[9] His remark, however, assures us that they were in use at an earlier date.

Astrolabes

The astrolabe was the astronomical instrument *par excellence* of the Middle Ages. It is constructed by stereographic projection, whereby points on a sphere are transferred to a plane surface. Figure 10.4a shows the principle of North Stereographic Projection. The sphere with centre O and south pole S is bisected centrally by a horizontal plane; points A and B on the sphere are transferred to points A_1 and B_1 on the plane. It can be demonstrated that angular relationships between points (and hence also lines) on the sphere remain unaltered by the transfer to the plane.

The essential parts of the instrument are the plate, the body, the rete and the alidade. The plate (Figure 10.4b) consists of a metal disc marked out by stereographic projection for the observer's latitude; it

Figure 10.4: Astrolabe

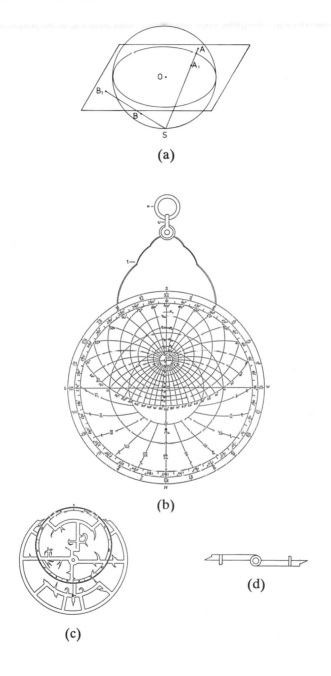

(a)

(b)

(c)

(d)

shows his meridian and zenith, arcs or circles of equal altitudes including the horizon and, radiating from the zenith, lines of azimuth (i.e. bearings). Around the centre of the plate are circles for the Tropic of Cancer, the Equator and the Tropic of Capricorn, the last named being coincident with the rim of the plate. It is usual to add a 'crepuscular' line for the time of twilight outside the horizon (not shown on Figure 10.4b). There is a hole in the centre of the plate. The body is also a circular metal plate with a hole in its centre. It is surrounded by a raised annulus, the outside of which is divided into quadrants, each subdivided into degrees. Inside this ring are two semicircles divided into twelve hourly divisions. At the top of the body is an attachment with a hole in it, through which the suspension ring passes (Figure 10.4b). The back of the body can carry various devices, not absolutely essential to the astronomical uses of the instrument. These usually include the 'shadow squares' for measuring terrestrial objects, as described in Chapter 7, and arcs for the unequal hours.

The rete, which rotates over the plate, is made of open metalwork, so that the lines on the plate are visible through it. It is essentially a star map and has holes (or sometimes gemstones) in it to represent the major fixed stars, each at its correct right ascension and declination. An eccentric ring represents the ecliptic; this is divided into the twelve signs of the Zodiac, each subdivided into 30 degrees. There is a hole in the centre of the rete (Figure 10.4c). The alidade is commonly of the shape shown in Figure 10.4d, and usually carries sights. In most cases, an alidade is also provided for the back of the astrolabe. The instrument is assembled by placing the plate on the body, inside the annulus, then placing the rete over the plate, then the alidade over the rete. The other alidade, if one is provided, is placed in position on the back. A pin is then passed through all the parts, and secured behind the rear alidade by a cotter.

A number of problems can be solved directly with the astrolabe, without recourse to calculation. To take but two examples:

1. To find the time of sunrise on June 2 (June 2 = 12 Gemini). The rete is rotated until the point 12 Gemini on the edge of the Zodiac circle comes up to touch the eastern horizon line on the plate below. The rete is held in this position and the rule is laid across the point. The time is read off where the end of the alidade nearest 12 Gemini crosses the hour scale.

2. To find the time of rising of a fixed star and its bearing as it rises

on (say) February 13 (February 13 = 25 Aquarius). The rete is rotated until the point representing the star rises to cross the horizon line. With the rete in this position the alidade is laid across 25 Aquarius; the time is read from the hour scale, and the azimuth under the point representing the star.[10]

These results, and many others, can be obtained in a few seconds, whereas considerably longer time would be needed to obtain them by calculation. Moreover, the use of the astrolabe can be mastered quite quickly by anyone with a rudimentary knowledge of spherical astronomy. These advantages, coupled with the capability of the instrument for measuring altitudes, explain its popularity from Hellenistic times until the eighteenth century. Because astronomers prepared tables of quantities to be used in the construction of astrolabes, any competent craftsman with a set of such tables at his disposal could manufacture astrolabes. Even so, the most sought-after instruments were those constructed by astronomers themselves.

The origins of the astrolabe can be firmly placed in the school of Alexandria. It was almost certainly known to Ptolemy and was described by Theon of Alexandria (*c.* 350), whose writings are preserved in the treatise of Serverus Sebokht, composed in Egypt before 660, i.e. a few years after the Arab occupation of the country.[11] The earliest Arabic treatises are those of Messahalla (d. *c.* 815), 'Alī b. 'Īsa (*fl. c.* 830) and Muhammad b. Mūsà al-Khuwārizmī (d. *c.* 835). The earliest Islamic instruments preserved date from the second half of the tenth century.[12]

The standard astrolabe has one disadvantage, in that it can only be used for one latitude. The problem can be partly overcome by having several plates for different latitudes and inserting the appropriate plate for a given location. This inconvenience was remedied by the Spanish Arab al-Zarqālī (Azarquiel, Arzachel) who made the vernal or the autumnal point the centre, and the meridian passing through the solstitial points the plane of projection. In its final form, which al-Zarqālī called *al-abbadiyya* in honour of Mu'tamid b. 'Abbād, king of Seville (1068-91), the entire instrument consists of a single tablet with two small subsidiary pieces. On the face of the tablet in stereographical 'horizontal' (as opposed to the ordinary 'vertical') projection the equator is represented with its parallels and its circles of declination, and the ecliptic with its circles of latitude and longitude; the projections of the equator and the ecliptic, then, are straight lines through the centre. The tablet is valid for any geographical latitude.

Moreover, since the projections of the two hemispheres exactly coincide, it suffices to add the principal stars to make it replace the rete of an ordinary astrolabe. Through the *Libros del Saber*, a compendium of Arabic knowledge compiled in 1277 in Spanish under the auspices of Alfonso X of Castile,[13] it became famous in Europe under the name of *Saphaea* (a corruption of the Arabic word for 'plate').[14] The ordinary astrolabe remained, however, by far the commonest astronomical instrument in general use during the Middle Ages.

It has now been established that the first European treatises on the astrolabe were of Arabic inspiration, and were produced by a certain Llobet or Lupitus and his colleagues at the close of the tenth and the beginning of the eleventh century, in the abbey of Ripoll in Catalonia.[15] The abbey was visited by Gerbert d'Aurillac, later Pope Silvester II, about 967, and it is probably due to him that the new knowledge was diffused into Europe. Most of the early Latin texts are anonymous or of doubtful attribution, with the exception of the treatise of Lupitus and one attributed to Hermann le Boiteux (d. 1054), abbot of Reichenau. All of them are very elementary, deal with only the simplest problems and contain many mistakes. Poulle argues convincingly that the reason for this lack of understanding was the absence of 'demonstration models':

Un traité de l'astrolabe, que ce soit dans la littérature arabe ou dans la littérature latine, ne peut se concevoir que comme la justification (traités de construction) ou le mode d'emploi (traités d'usages) d'un instrument que le lecteur pouvait manier, démonter ou manoevrer à son aise au fur et à mesure de la lecture. Or on ne connait pas d'astrolabe latin in cette époque. Cela ne prouve évidemment pas qu'il n'en ait pas existé; mais cela n'est tout de même pas certain. On peut se demander si, faute d'instruments latins, les milieux scientifiques occidentaux n'auraient pas alors disposé seulement d'un matériel arabe: on s'expliquerait par conséquent ainsi les difficultés qu'ils auraient rencontrées pour décrire ou pour utiliser des instruments dont les graduations et les inscriptions devaient rester obscures à la plupart.[16]

These texts are also full of unassimilated Arabic words, sometimes so distorted as to make the meaning obscure. During the twelfth century many translations from Arabic works were made in the archiepiscopal city of Toledo. Among many astronomical works, the *Planisphere*, or theory of stereographic projection, by Ptolemy was translated by Hermann of Dalmatia, and the treatise on the astrolabe by Messahalla was translated by Jean of Seville. These, and several other translations from Arabic into Latin on the construction and uses of astrolabes, stimulated the production of a number of original treatises by European writers. These included: Raymond of Marseilles, before 1141, on the construction and uses of the astrolabe; Adelard of Bath, about 1142-6, on its construction; Robert of Chester, in 1147, on its uses; Abraham ibn Ezra, about 1158-61, on its uses. These treatises are much better than their predecessors of the tenth and eleventh centuries. Arabic words no longer appear as equivalents of Latin expressions, and the Latin terms themselves are those that were to be used from that time onwards in the definitive technical vocabulary of the astrolabe. The twelfth-century treatises, in their discussion of the uses of the instrument, reveal an awareness of its resources, and present problems in a logical sequence and in ascending order of difficulty. The geometric constructions on the back of the astrolabe, including the marking of the ancient quadrant, were known in Europe by the tenth century, but are first described clearly by Raymond of Marseilles, and came into general use in the thirteenth century.[17]

Equatoria

During the late Middle Ages a number of computing instruments were devised having as their common objective the determination of the longitude of any one of the planets for any given time. This was done by constructing to scale and by mechanical and graphical means the Ptolemaic configuration for that particular planet at the given instant. The desired result was then simply read off a scale of the instrument, and, being precise to within a degree, was useful for, say, casting horoscopes. Where great accuracy was not required these equatoria, as they were called, relieved the tedium of numerical computation.[18]

The basic essentials for the construction of equatoria are described by North as follows:

The first planetary equatoria were, in effect, nothing more than movable representations, to scale, of movements around the epicycle and deferent circle of Ptolemaic theory. Discs representing the epicycle and deferent circle of any particular planet, such that their radii were in the correct ratio, could clearly be moved within a zodiac circle so that, perhaps with the assistance of threads of rules, they were correctly placed for a given moment of time. A point on the epicycle disc will now represent the planet, while, laying a rule or stretching a thread through this point and the centre representing the centre of the Earth, the final ecliptic longitude will be indicated at the zodiac scale. The correct placings of the epicycle will depend upon a knowledge of the mean motions around the deferent circle and around the epicycle (i.e. in argument), as well as on a knowledge of the eccentricity (i.e. the position of the Earth and of the equant, in relation to the deferent centre). These mean motions will in turn be found from tables, or their equivalent. As we have so often seen, the work *motus* of medieval Latin denotes most commonly a position, or a change in position, rather than a motion. To discover the mean motus, or mean positions in the two basic circles, we must also know the *radices* of the motions, this is to say, the values of the angles at some standard epoch. Once this information is available, however, it is no more difficult to place the two discs than to draw the appropriate diagram of *Almagest* to scale, for the moment in question, and to find thereby the planetary longitude. What matters most is that by solving the problem in this way there is no longer any need to introduce planetary equations (equation of the centre and equation of the argument). Accuracy will inevitably be lost, with an instrument or drawing of normal dimensions, but the method is simple and rapid, and it greatly appealed to the mathematically hesitant Middle Ages.[19]

Physically, as the above description indicates, the instrument, for a given planet, closely resembles the Ptolemaic diagram shown in Figure 10.1. The main plate was graduated, along its outside edge, into 360 degrees and the signs of the Zodiac. The deferent circle and the graduated equant circle were engraved on the surface of the plate. A small epicycle plate, its outer edge graduated into 360 degrees, was provided for the planet. There were small holes in this plate to take a thread, and a 'window' through which the equant scale could be viewed. A thread from a peg in the equant centre was passed through the small holes in the epicycle plate, and another thread, from a hole in the centre of the main plate, was stretched over the epicycle plate

and the zodiac scale. After the necessary values had been obtained
from tables, the threads and the epicycle plate could be manipulated
to give the planetary longitude for a given moment.

The early treatise on equatoria are shown in Table 10.1.[20]

Table 10.1: Various Early Treatises (Equatoria)

Name	Date	Location
1. Ibn al-Samh	*c.* 1025	Spain
2. Azarquiel	*c.* 1050	Spain
3. Abu al-Salt	*c.* 1110	Spain, Egypt and Tunisia
4. Campanus of Novara	1264	Italy
5. Richard of Wallingford	1326	England
6. John of Linieres	*c.* 1330	France
7. The Merton Instrument	*c.* 1350	England
8. Chaucer	*c.* 1392	England
9. Jean Fusoris	1414	France
10. al-Kashi	1416	Iran and Turkestan

The treatises of Ibn al-Samh and Azarquiel found their way, in
Castilian, into the *Libros del Saber,* but there is no evidence of their
having been translated into Latin. The intricate structure of the
equatorium of Campanus is different from those of his predecessors,
although all are based firmly upon Ptolemaic theory. Although
diffusion from Islam remains the most likely explanation for the
origin of European equatoria, it is possible that it was an independent
invention. The essential idea is much more obvious than the principle
of the astrolabe, and could have been derived directly from treatises
on Ptolemaic astronomy. *A Theorica planetarum,* for example,
was very frequently copied in the thirteenth century and after, and
is especially noteworthy for the number of manuscript copies having
threads in the diagrams.[21]

These instruments, which are the most important of the medieval
astronomical devices, are significant for several reasons. One is their
common origin, in concept or reality or both, in the Eastern Mediterranean in Hellenistic times, their diffusion to and development in
Islam, and their adoption in the West — in the case of astrolabes and
equatoria at least — about the end of the tenth century. They also
exhibit a readiness and a capability for converting astronomical
relationships into mechanical reality for computational purposes.
They are clearly of importance, therefore, in the history of analogue

computers of all kinds. Many of the techniques and components incorporated in them were later to be part of the array of ideas that were embodied in the mechanical clock, and into mechanical technology as a whole.

Notes

1. Derek de Solla Price, *Science since Babylon*, enlarged edn (Yale University Press, New Haven and London, 1979), p. 50.

2. Richard of Wallingford, *Life and Works*, ed. J.D. North (3 vols., Oxford University Press, Oxford, 1976), vol. 3, pp. 168-200. In these three splendid volumes, North presents the works of Richard of Wallingford in the original Latin, with translations and commentaries. The present reference is to Appendix 29 — 'An outline of the Ptolemaic theory of planetary longitude, as applied in the Middle Ages'.

3. A.I. Sabra, ' 'Ilm al-Hisāb', in EI, vol. 3, pp. 1138-41.

4. D. Pingree, ' 'Ilm al-Hay'a' in EI, vol. 3, pp. 1135-38.

5. David A. King, 'On the Astronomical Tables of the Islamic Middle Ages', *Colloquia Copernicana*, III (1975), pp. 37-56.

6. Derek de Solla Price, *Gears from the Greeks* (Science History Publications, New York, 1975), *passim.*

7. Price, *Science since Babylon*, p. 48.

8. E. Wiedemann, 'Ein Instrument, das die Bewegung von Sonne und Mond darstellt, nach al-Bīrūnī', *Der Islam*, vol. 4 (1913), pp. 5-13.

9. Price, *Gears from the Greeks*, p. 54.

10. Henri Michel, *Traité de l'Astrolabe* (Alain Brieux, Paris, 1976); comprehensive treatment of the subject, giving history, construction and use of all types of astrolabe. For Islamic astrolabes see W. Hartner, 'Asturlāb' in EI, vol. 1, pp. 722-8. Working models with explanatory booklets are available, e.g. H.N. Saunders, *The Astrolabe*, Bude, 1971 (astrolabe in cardboard and plastic), and R.S. Webster, P.R. MacAlister and F. M. Etting, *The Astrolabe: Some Notes on its History, Construction and Use* (astrolabe in cardboard) (Lake Bluff, Illinois, 1974).

11. O. Neugebauer, 'The Early History of the Astrolabe', *ISIS*, vol 40 (1949), pp. 240-56.

12. Hartner in EI, Vol. 1, pp. 722-3.

13. *Libros del Saber de Astronomia*, M. Rico (ed.) (Madrid, 1863).

14. Hartner in EI, vol. 1, p. 726.

15. Emmanuel Poulle, 'Les instruments astronomiques de l'Occident latin aux XIe et XIIe siècles, *Cahiers de civilisation medievale*, vol. 15 (1972), pp. 27-40. *Idem*, 'L'Astrolabe Medieval', *Bibliothèque de l'École de Chartres*, no. 112 (1954), pp. 81-103.

16. E. Poulle, 'Les instruments astronomiques', p. 32.

17. Ibid., p. 35.

18. E.S. Kennedy, 'The Equatorium of Abu al-Salt', *Physis*, Year XII, Fasc. 1, Florence (1970), pp. 73-81, p. 73.

19. Richard of Wallingford. *Life and Works*, vol. 2, pp. 250-1.

20. Kennedy, 'The Equatorium of Abu al-Salt', p. 80.

21. Richard of Wallingford, *Life and Works*, p. 258.

11 Automata

Professor Price has argued strongly and consistently that mechanistic philosophy led to mechanism — the urge to represent the universe by mechanical means — rather than the other way about. He predicates 'a deep-rooted urge of man to simulate the world about him through the graphic and plastic arts'. A natural extension of this urge is to impart movement to static simulations and so create automata.[1] The weight of the evidence seems to favour this hypothesis, which assigns a key role to the makers of automata in the development of a rational, mechanistic view of the universe. Certainly, the theme of man-made organisms, magically endowed with a life of their own, has exercised a powerful fascination since the early Greek legends such as that of Pygmalion, the animated beasts of the Thousand and One Nights, the miraculous animation of holy images, the doll Olympia in the Tales of Hoffman and the monster of Frankenstein, to take but a few examples. The same fascination has persisted into modern times, with the sinister electronic robots of science-fiction films. The urge to re-create the phenomena of the visible world did not manifest itself only in the construction of animated organisms, but also in the production of celestial automata. Indeed, the two kinds of mechanical representation often occurred in a single machine.

Nevertheless, it would be wrong to invest all automata with a special significance in the development of man's view of the world. Many were simply ingenious toys, designed to amuse or mystify, or devices such as fountains for giving aesthetic pleasure. For this reason, some historians, far from detecting any significance in these devices, have dismissed them as trivial. This is about as sensible as dismissing modern communications technology because many television programmes are frivolous or banal. The comparison is not far-fetched: as is the case with astronomical instruments, many of the ideas developed in the construction of ingenious devices were later to enter the general vocabulary of mechanical technology. It is a little unfortunate that the term 'automata' is used to designate this type of

199

construction, if only because, from an engineering point of view, it is usually the activating mechanisms, rather than the displays themselves, that embody the most interesting ideas. On water-clocks and early mechanical clocks, time intervals were signalled by the activation of automata; since these were an integral part of the clocks, they will be described in Chapter 12. Here we are concerned with a range of devices, some with a serious purpose, many designed to amuse, startle or perplex the onlookers.

The prehistory of automata begins with the dolls with jointed arms and other articulated figurines such as those from the ancient Egyptian tombs from the Twelfth Dynasty onwards. The next stage of development is also found in ancient Egypt: talking statues worked by means of a speaking trumpet concealed in hollows leading down from the mouth. Two such statues are extant: a painted wooden head of the jackal God of the Dead is preserved in the Louvre, and a large white limestone bust of the god Re-Hamarkhis of Lower Egypt is in the Cairo Museum. The articulated masks to be worn over the face, found in Africa and elsewhere, probably served a similar purpose.[2] The beginnings of sophisticated, mechanised automata, however, are to be found in Hellenistic Egypt. Before the later Middle Ages, because of their fragility, nothing remains of any of the early devices, and we have to rely for our information on passing references in literary works, and upon a few precious treatises. The devices to be discussed here did not include celestial representations; not surprisingly, these are confined to astronomical instruments and to clocks, since time and the heavens were considered as a single idea. Nor did they always incorporate biological simulacra, particularly in those devices, such as trick vessels and fountains, where the machine itself *was* the automaton.

The first major advances in this branch of fine technology seem to have been made by Archimedes and by an engineer called Ctesibius, who worked in Alexandria about 250 BC. His work is known to us only through Vitruvius, who described how Ctesibius, the son of a barber, devised a mechanism for raising and lowering a heavy mirror in his father's shop. The mechanism included weights that moved up and down in cylinders, and Ctesibius discovered that they would move more easily if holes were made in the bottoms of the cylinders. Vitruvius continues:

Thus when the weight ran down into the narrow tubes, and compressed the air, the large amount of air was condensed as it ran violently down through the mouth of the tube and was forced into the open; meeting with an obstacle, the air was produced as a clear sound.

Ctesibius, therefore, when he observed that the air being drawn along and forced out gave rise to wind-pressure and vocal sounds, was the first to use these principles and make hydraulic machines. He also described the use of water-power in making automata and many other curiosities, and among them the construction of water-clocks.[3]

It appears, therefore, that Ctesibius is to be credited with the invention of the organ. His contributions to the development of the water-clock are discussed in the next chapter.

One of the most important names in the history of fine technology is that of Philo of Byzantium, who flourished about 250 BC. He is mentioned by Vitruvius as a constructor of machines,[4] and by the Egyptian historian al-Kindī (AD 897–961), who says that he built water-wheels, mills and ingenious devices (*hiyal* in Arabic).[5] (He is thus but one example, among several others, of an engineer who constructed both utilitarian machines and automata.) Apart from a fragment in Latin, the *Pneumatics* of Philo exists only in Arabic versions. These may contain some later Islamic interpolations, but for the most part the descriptions of ingenious devices can be accepted as authentic.[6]

Another important surviving text is the *Pneumatics* of Hero of Alexandria (*fl.* mid-first century AD).[7] Hero's works were well known to the Arabs, both the *Pneumatics* and the *Mechanics*, the latter existing only in an Arabic version made by Qustā b. Lūqā in Baghdad in the ninth century.[8] He is also said to have written a work on water-clocks, but this treatise has not come down to us.[9] For the most part, Hero's designs, in the *Pneumatics*, are modelled closely upon those of Philo, but there are also some original devices of considerable interest. From the same period as Hero we have descriptions by Tacitus and Suetonius of Nero's 'Golden House', which, with its grounds, occupied at least 125 acres of land inside the city bounds of Rome. In the words of Suetonius:

He built a palace stretching from the Palatine to the Esquiline, which at first was called the Domus Transitoria, but after it had been burnt and restored was called the Domus Aurea. The following facts will serve to indicate something of its size and luxury. In the vestibule there was a colossal statue of Nero, 120 feet high; and so spacious was it that it had a triple portico a mile long. There was an artificial lake to represent the sea, and on its shores buildings laid out as cities; and there were stretches of countryside, with fields and vineyards, pastures and woodlands, and among them herds of domestic animals and all sorts of wild beasts. Elsewhere all was overlaid with gold, and bright with jewels and mother-of-pearl. There were dining-halls whose coffered ivory ceilings were set with pipes to sprinkle the guests with flowers and perfume. The main dining-hall was circular, and it revolved constantly, day and night, like the universe.[10]

No doubt the talents of engineers such as Hero were called upon to provide the marvels in palaces such as this.

Islamic engineers drew their initial inspiration from treatises such as those of Philo and Hero but as we shall see a little later, they developed more advanced methods of automatic control. Indeed, in the field of machines operated by small differences in aerostatic and hydrostatic pressure, the work of the Banū Mūsà was probably unsurpassed until modern times. These three brothers were called, in order of seniority, Muhammad, Ahmad and al-Hasan. Their father, Mūsà b. Shākir, was a noted astronomer who was a close companion of the future Caliph al-Ma'mūn when the latter was living in Khurasan. After Mūsà's death, the brothers were entrusted to the care of al-Ma'mūn (reigned 813–33) and became members of the courtly circles of this Caliph and his successors. They took part in the turbulent palace politics of the time, undertook major public works, and extended their patronage to a number of eminent scientists and translators. They also made journeys to Byzantium, and were instrumental in bringing back to Baghdad copies of scientific works, which were then translated from Greek or Syriac into Arabic. They were also eminent scientists in their own right, and composed a number of works on astronomy and mathematics, one of which, *On the Measurement of Plane and Spherical Figures*, was translated into Latin by Gerard of Cremona in the twelfth century. There are entries on the Banū Mūsà in all the major Arabic biographical dictionaries, and their political activities are mentioned by the great historian al-Tabarī (839–923), who was alive when they were prominent in public affairs.[11] Their *Book of Ingenious Devices*, containing descriptions of 100 devices, was probably mainly the work of Ahmad. It was widely appreciated in medieval Islam. Ibn Khaldūn says of it: 'There exists a book that mentions every astonishing remarkable and nice mechanical contrivance. It is often difficult to understand, because the geometrical proofs occurring in it are difficult. People have copies of it. They ascribe it to the Banū Shākir.'[12]

By the end of the tenth century, the construction of automata was probably a well-established practice, since part of a scientific encyclopaedia is devoted to the subject. This is *The Keys of the Sciences (Mafātīh al-'Ulūm)*, compiled by Abu 'Abd Allah al-Khuwārizmī about AD 980. The Eighth Treatise deals with ingenious devices (*hiyal*), and lists a number of components and techniques, with etymological information, that were used by makers of these machines. This is a particularly useful work, since al-Khuwārizmī does not limit himself to definitions, but includes descriptions of manufacturing processes.[13]

A most important treatise was written in Muslim Spain in the eleventh century by a certain al-Murādī. Unfortunately, the only known manuscript copy is so badly defaced that it is impossible to deduce from it precisely how any of the machines were constructed. Most of the devices were water-clocks, but the first five were large automata machines that incorporated several significant features. Each of these devices, for example, was driven by a full-size water-wheel, a method that was employed in China at the same period to drive the monumental water-clock of Su Sung. The automata were of the type that we shall meet with frequently when we discuss water-clocks, for example a set of doors in a row that open at successive intervals to reveal jackwork figures. The text mentions both segmental and epicyclic gears, and although the illustrations are in other respects quite incomprehensible, they clearly show gear-trains incorporating both types of gearing. This is extremely important: we have already encountered segmental gears in al-Jazarī's water-raising machines, but here we have, over a century earlier, segmental and epicycle gears used to transmit high torque. Nothing approaching this sophistication in the use of gears appears in Europe before the middle of the fourteenth century. Another feature of al-Murādī's machines is his use of mercury in some kind of balancing system. As we shall see, mercury was used in several types of Islamic devices. Al-Murādī's treatise is clearly a document of great significance, and it is to be hoped that a better copy comes to light some day.[14]

The treatise on machines by al-Jazarī, completed in Diyar Bakr in 1206, has already been mentioned in Chapter 8. We know nothing of his life beyond what he tells us in the introduction to his book. At the time of writing, he was in the service of Nasīr al-Dīn, the Artuqid prince of Diyar Bakr. He had by then spent 25 years in the service of the family, having served the father and brother of Nasīr al-Dīn before him. He was ordered to write the book by his master, because many of his devices were fragile and knowledge of them would otherwise be lost to posterity. We therefore owe to this prince of an unimportant dynasty, a vassal of Saladin, one of the most important engineering documents to have been written in any cultural area before modern times. Its importance lies not only in the machines, and the techniques and components incorporated in them, but in the method of presentation. Each of the 50 devices is described and illustrated in scrupulous detail, with the avowed (and successful) intent of enabling later craftsmen to reconstruct the machines. The approach is very unusual — almost unique. Most craftsmen did not commit their results to paper, usually because they were illiterate, sometimes because they did not wish to disclose the secrets of their craft. Treatises were often written by laymen who did not really understand the technical side of the subject (e.g. Frontinus), and hence their descriptions tend to be in broad general terms. Al-Jazarī's work, however, enables us to know, not just the machines in their finished form, but how their parts were manufactured, assembled and interconnected.[15] The number of surviving manuscripts, copied between the thirteenth and the eighteenth centuries, attest to the continuing interest in his work in the Muslim world. Although most of the major biographical works had been written before his time, he is mentioned by the Egyptian scholar al-Qalqashandī (1355–1418).[16] No translation of his work, into Latin or any other European language, is known before modern times.

Of the six categories into which his book is divided, the first, on water-clocks and candle-clocks, is perhaps the most important; these clocks, which include arrays of biological and celestial automata, will be discussed in the next chapter. Of the remaining categories, three are of direct relevance to the present subject. These are: II — trick vessels; III — phlebotomy measuring devices and water dispensers; IV — fountains and musical automata. Apart from the high engineering content, it must be said that in categories II and III, al-Jazarī's devices do not show any advance on similar devices described by the Banū Mūsà. He tends to rely rather more on mechanical methods, rather than the small pressure differentials used with such skill and delicacy by his predecessors. Indeed, the brothers may be said to have exhausted the possibilities of obtaining results by contriving small variations in aerostatic and hydrostatic pressures. On the other hand, al-Jazarī's fountains and musical automata are considerably more advanced than those of the Banū Mūsà; he seems to have been more at home with fairly large machines.

There are a number of Arabic manuscripts which contain material from a number of sources. Thus in one manuscript there may be excerpts from Philo's *Pneumatics,* one or two devices from the Banū Mūsà and al-Jazarī, together with descriptions of machines of unknown origin. Borrowings from the Banū Mūsà and al-Jazarī are not acknowledged as such, and the descriptions are different from and inferior to the originals. One such manuscript is in the Bodleian Library, Codex Oxford 954. It contains, as do other manuscripts of this type, several perpetual motion machines — 'wheels that move by themselves'. Although the idea is a physical impossibility, the machines are not wholly absurd. Two of them (6r–7r and 8v–9r) are intended as water-raising machines, and have wheels containing mercury as part of the mechanisms. We know that such wheels could be used as effective clock escapements (see Chapter 12), and it may well be that the driving weights were deliberately omitted in order to keep the operation secret. Oddly enough, another machine (7v–8r) is in fact a weight-driven pump, in which two outer gear-wheels driven by lead weights mesh with a large central gear-wheel — at least this seems to be the gist of the rather obscure text and illustration. This machine could probably have been put to practical use. Machines of this type, in various manuscripts, might repay close study, although it must be said that obscurities in the texts make this a formidable task.

There is evidence that the skills of automata makers were enlisted to add distinctive features to royal palaces, as had been the case in Nero's Golden House in Rome. In 917–18 the writer Khātib described the wonders of the newly-built palace of the Caliph al-Muqtadir in Baghdad. These included a tree of silver, in a large pond, with 18 branches and multiple twigs, with silver or gilt birds which whistled at times. On both sides of the pond were 15 statues of mounted horsemen which moved in one direction as if chasing each other. There was a mercury pond, 30 cubits by 20, with four gilt boats and around it was a fabulous garden.[17] A little later we have a description in a Byzantine treatise of the 'Throne of Solomon' in Constantinople. The automata surrounding this seat of the emperor included a tree with singing birds, roaring lions and moving beasts. These details are confirmed by Liutprand of Cremona, who visited Byzantium as envoy in 948 and again in 966. These automata could have been derived direct from the works of Hero or could have been diffused from Baghdad, where in any case the tradition was based upon the writings of Hero and Philo.[18] Trees with singing birds were a common motif in the automata-treatises of Hellenistic engineers. Decorative fountains were used to enhance the attractions of formal gardens and courtyards. In the tenth century, al-Muqaddasī remarked upon the beauty of the fountains in Damascus,[19] and about two centuries later, in the same city, Ibn Jubayr saw a fountain consisting of a single jet surrounded by small pipes that threw the water up like the branches of a tree.[20]

The early history of automata in Europe takes the form of legendary tales that probably reflect some hazy knowledge of Hellenistic/ Arabic automata. Into this category come the extraordinary stories which attribute magical powers to the poet Virgil. This legendary material is, of course, of quite a different nature to the veneration of Virgil as a great poet, which caused Dante to assign to him a key position in the *Divine Comedy*. The earliest work to include such tales is *Policraticus*, written about 1159 by John of Salisbury. Over the next two centuries, various works, nearly all of north European origin, added to and embroidered the legendary tales. Virgil is said to have been able, for example, to build a bridge of air to take him anywhere. In a story that may date back to the eighth century, he is said to have built a noble palace in Rome in which there were wooden statues, each representing a province of the Empire. When trouble arose in a province, its representative rang the bell, and a bronze horseman on top of the palace brandished his spear and pointed it towards the province that was in difficulties.[21] In a similar vein are the stories that endow philosophers with magical powers. Albertus Magnus (d. 1280), for example, is said to have made a brazen head which whispered secrets to him.[22]

Apart from such legends, there is not much evidence for the manufacture of automata in Europe until late in the Middle Ages, when jackwork figures appeared on water-clocks and then on mechanical clocks. In the Notebooks of Villard de Honnecourt (*c.* 1254) there is a simple rope-and-pulley apparatus for turning the figure of an angel, and a rope-driven automaton bird.[23] These are very crude, however, when compared to the sophisticated devices of Hero or the Banū Mūsà. Nevertheless, there seems to have been a medieval tradition, probably continued from Roman times, for the construction of waterworks. A pair of beautiful Norman drawings of the waterworks of Canterbury Cathedral and its vicinity, indicates that hydraulic engineering was in the hands of able craftsmen by about AD 1165.[24] Although one or two of al-Jazari's devices are intended to play mild practical jokes, he constructed nothing like the famous pleasure garden built for Duc Phillippe, Count of Artois, at his castle of Hesdin towards the close of the thirteenth century. This contained hidden spouts for wetting ladies from below and covering the company with soot and flour, along with a large number of animated apes covered with real hair and sufficiently complicated to need frequent repair.[25] We have no descriptions of the method of activation of these devices; for an understanding of the construction and operation of automata we must return to the works of the Hellenistic and Islamic engineers.

Components and Techniques

Most of our information about manufacturing and constructional methods comes to us from al-Jazari's work. This is because he gives us step-by-step instructions, describing the manufacture of a tank, for example, from the sheet of copper to the finished vessel. Other writers simply say 'A tank of 2 spans diameter, 5 spans long, is installed ...' Although al-Jazari was a better engineer than most, it is reasonable to suppose that his predecessors' methods were similar to his, except where he makes it clear that a particular technique is of his own devising. Apart from some specialised water machinery, the components and techniques described in this section are also applicable to the construction of water-clocks. The subject can be conveniently divided into two categories — mechanical and hydraulic.

Mechanical

Wheels, Axles and Bearings. Wheels were nearly always pulleys, the small ones made of copper, the large ones of wood, both having semicircular grooves in their rims to accommodate the ropes or cords. Large wheels were sometimes made of laminated timber to minimise warping and, if necessary, they were statically balanced; they were placed on a mandrel and checked for free rotation; if they were out of balance, small pieces of lead were fixed to the outer edge until balance was achieved. Quite often, especially in clocks, large pulley-wheels had to be precisely dimensioned, so that their rotation converted the travel of a float, for example, into a required circular or linear distance. Wheels were usually fixed rigidly to their axles. These could be of wood, copper or iron, depending on the duty. Wooden axles had iron 'acorns' nailed to their ends and these protuberances were mounted in iron journals. In the terminology, distinctions were made between large and small horizontal bearings, and between these and vertical thrust bearings. There is no mention anywhere in the literature of lubrication.

Gears. All the combinations of gears appear in the automata treatises: parallel meshing; meshing at right angles; worm-and-pinion; rack-and-pinion. As we have noted, segmental gears appear in al-Murādī and al-Jazarī and epicycle gears only in al-Murādī. Larger wheels were usually made of wood, the teeth formed from pegs fixed around the perimeter; for parallel meshing one of the gears was often a lantern pinion. Smaller gears were made from copper or bronze, the teeth filed to the shape of equilateral triangles. In automata and water-clocks gears were rarely used in the main transmission system; their use was usually confined to peripheral mechanisms — pulley-trains were commonly used for transmitting power. Al-Murādī's treatise has therefore three unique features: epicycle gears, the use of gears to transmit power, and the fact that the gear-wheels were all made of metal.

Figures. The material most commonly used for making the figures of humans, animals and birds was beaten copper. If the limbs of the head were required to move they were attached to the torso by hinges or pin-joints. Figures were also made from beaten brass and from wood; if the material was wood, parts were hollowed out only if the mechanisms were to be placed inside them. When lightness was essential al-Jazarī made his figures from papier mâché, but this material does not occur in the works of earlier writers. In the treatises of Philo, Hero and the Banū Mūsà the figures seldom make bodily movements; they are either static, or the whole figure moves rotationally or in a straight line. Many of al-Jazarī's figures, however, move parts of their bodies. Human figures turn their heads, raise their arms or move their legs, birds spread their wings, and so on. It is rather surprising that the medieval tales of animated man-made beings could have grown out of the rather passive simulacra of al-Jazarī's predecessors.

Fixings and Fittings. There were no wood-screws or nuts-and-bolts until after the end of the Middle Ages. Although mortise-and-tenon and dovetail joints had been known since ancient times, these are not mentioned by the automata-makers, who used nails for joining timbers. Metal pieces were joined end to end by male/female joints (the same expression was used then). When several pieces were on the same axle, they were held in place by cotters — as in the astrolabe. The usual method of making rigid joins between two pieces of metal was by soldering. Hinges were of the pin-jointed type, but one-way hinges that could flex only in one direction found a special application in clockmaking. Ropes and cords were made from hemp or silk, chains from iron or copper, and wire usually from copper.

Hydraulic Components

Pipes, Channels and Siphons. Pipes were made by bending a sheet of copper round a cylindrical wooden former, and then soldering along the seam. Channels, either of semicircular or rectangular cross-section were also usually made from copper, occasionally from brass. Both bent-tube and concentric siphons were widely used. The latter, which works in exactly the same way as the ordinary bent-tube siphon, is shown in Figure 11.1a. Pipe (a-cc) was probably joined to pipe (bd) by copper wires soldered to each pipe. Pipe (bd) passes through the floor of the container, the joint being soldered to an air-tight fit. When liquid is poured into the container it rises to point (b), and the discharge of the first drop of water reduces the air pressure in space (ab) sufficiently to cause a continuous flow through pipe (bd) until the liquid level reaches (cc).

Figure 11.1b shows a double concentric siphon, which was used only by the Banū Mūsà; a closed pipe is fitted to the lower end of the narrow pipe as well as over its top. Usually this device was fitted into

Figure 11.1a: Concentric Siphon

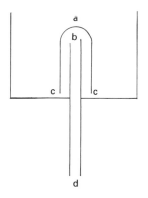

Figure 11.1b: Double Concentric Siphon

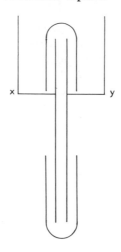

the neck of a pitcher with a narrow spout — in Figure 11.1b, (xy) is a plate soldered to the circumference of the bottom of the pitcher's neck. Unfortunately, a mathematical analysis of the way this apparently simple device works is fairly complex, and cannot be discussed here.[26] The effects of introducing it are as follows: if the neck is open to atmosphere, and a liquid is poured in, the liquid will flow through the siphon until inpouring is stopped. When inpouring is started again, the siphon will not be re-activated until the liquid in the neck reaches a certain critical level; if the neck is too short to allow this level to be reached, the liquid will simply spill out from the top of the neck. If the neck is made airtight, apart from a narrow filler tube in the closing plate soldered to the top of the neck, it will be impossible to recommence inpouring because the liquid head is insufficient to compress the air in the neck — the liquid will not enter the filler tube and will spill over the plate. If there is a concealed air-hole in the hollow handle of the pitcher, outpouring can be stopped by closing the air-hole with the finger.

Water-wheels. As mentioned in Chapter 9, horizontal axial flow wheels were made by cutting along the radii of metal discs and raising the segments to form curved vanes. Al-Jazarī uses scoop-wheels in some of his devices; these consist of a central hub from which a number of spokes radiate — at the end of each spoke is a ladle-like scoop into which the water is directed from a pipe. This primitive

Pelton wheel is also shown in one of Carra de Vaux's reconstructions of a machine in Philo's treatise, but there is no justification for this interpretation in the text, which seems to indicate a simple paddle-wheel.[27]

Vessels. Tanks and reservoirs, of rectangular or circular cross-section, were made from sheet copper, formed to shape and soldered along the seams. The large, vertical cylindrical reservoirs used in water-clocks were carefully made of uniform cross-section. They were made up of several drum-like pieces. Each drum was hammered around an exactly circular wooden template. A further check was then made by dividing the inside of the drum vertically into equidistant sections. Water was poured in until it reached the top of the first section; a similar quantity was then poured in between the first and second marks, and if the water-level coincided with the second mark all was well. If not, minor adjustments were made to the shape of the second section by hammering until the water-level coincided with the mark. And so on, to the end of the drum and for the other drums. When this had been done, the drums were placed together and joined by soldering copper rings around the joins to form the complete reservoir. All vessels that were to contain liquids were thoroughly tinned on the inside. Air-vessels were used for mechanical whistles. These were small vessels, airtight except for the entry pipe, the pipe to which the whistle was attached and a bent-tube siphon. The whistle was made, in the words of al-Jazarī, as follows:

Then one takes a piece of pipe 2/3 of a finger length long and so wide that it can be encircled by thumb and index finger — it is like a jar. At each end a cover is securely fitted. In the centre of one of the covers a hole is made shaped like the eye of a man, but smaller. Then a piece of copper is taken, as long as the third of an index finger, and as light as possible. This is hammered until its circular shape is changed into the shape of an eye. This is inserted into the hole until it is near the end of the jar, and is then securely soldered at the joint. The pipe is slightly narrower at the end inside the jar than it is at the point where it is soldered to the hole, and it should be as light as possible. Then a small hole is drilled in the centre of the other cover of the jar and on to this hole is soldered a pipe . . .[28]

The end of this second pipe is soldered to a hole in the cover of the air-vessel. When water runs into the air-vessel it covers the end of the siphon and air is expelled through the whistle. The water is evacuated from the air-vessel when it reaches the bend in the siphon. The description of the manufacture of the whistle is typical of al-Jazarī's detailed instruction for the manufacture of components — small or large.

Valves and Taps. Hinged clack-valves and plate-valves are used in the Hellenistic devices. The former is a circular plate hinged to one side of a pipe, and prevented from moving in one direction by a lug soldered to the wall of the pipe opposite the hinge. Plate-valves were placed over the inlet holes at the bottom of the vertical cylinders of pumps; vertical pins soldered to the bottom of the cylinder passed through holes in the plate and prevented it from moving out of place. On the suction stroke the plate was lifted to allow the water in, and fell back under its own weight when the pressure was released. Conical valves do not appear in the works of Philo or Hero, although one appears in a treatise on a water-clock attributed to Archimedes (see Chapter 12). Although they are used by Ridwān and al-Jazarī in water-clocks based upon the 'Archimedes' treatise, their most extensive application is in the Banū Mūsà's book. They used them with great confidence as automatic controls in hydraulic transmission system. (Conical valves first appear in Europe in the writings of Leonardo da Vinci.)[29] They were made of bronze, plug and seat being cast together and then ground together on the lathe, with emery powder, to a watertight fit. Taps were made in a similar manner. In all the treatises there are examples of taps with multiple borings: on the inlet side of the seat were a number of holes, each connected by a pipe to a different chamber in the vessel, each chamber containing a different liquid. There were the same number of holes in the plug, all discharging into the same outlet. A hole in the plug could be made to coincide with a given hole in the seat by turning the handle to a marked position. A number of different liquids could thus be drawn from the one tap.

Tipping-buckets. These devices, which were frequently used by al-Jazarī, were probably derived from the tipping-spoons that are used in one or two of Philo's machines. It consisted of a tapered vessel, like 'half a boat', closed by a vertical plate at the wide end. Stub-axles protruded from the top of the side plates, near the broad end, and these axles rested in journals on the side of a tank, the empty tipping-bucket being balanced horizontally. Water from an orifice dripped into the vessel and after a predetermined period — one hour say — the bucket tilted and discharged all its contents into the tank. They then ran out of the tank through a pipe and activated other mechanisms. In another application, there was a vertical rod extending above the top of the closing plate. When the bucket tipped, the end of this rod struck a projection on another mechanism and set it in motion. Tipping-buckets appeared in Europe as components of rain gauges in the seventeenth century, and were incorporated in the continuous recording gas calorimeter, perfected about 50 years ago.[30] We cannot tell whether these were re-inventions or whether the Islamic device was somehow transmitted to Europe.

Floats. These varied in size, from tiny ones with the plugs of conical valves soldered to their tops, to the very large ones used to drive water-clocks. The latter had a ring soldered to their upper surface to which the driving cord was attached and were weighted with sand or lead. The *tarjahār* was a submersible bowl, whose original purpose was to measure the periods for the allocation of irrigation water. It consisted of a hemispherical bowl, made of copper or brass, with a graduated orifice in its underside. It was timed to sink in a given period — half an hour or an hour — when it was placed on the surface of the water. Al-Jazarī adapted the tarjahar for use in a unique type of water-clock.

Orifices. The proper calibrations of orifices was essential for the accuracy of water-clocks. The bore of orifices was usually very small, 2 mm or less, and in order to get a uniform hole, not subject to rapid enlargement due to the action of the water, special materials were used. Ctesibius used gemstones of an unspecified type, whereas al-Jazarī always used onyx. It is assumed that he knew approximately the required size for a given orifice, because his method of calibration was to make a hole narrower than the minimum estimated bore, then enlarge it using copper wire and emery powder, checking the discharge rate at intervals, until the desired hourly rate of flow was reached. The orifice was then fitted to a brass housing, which was soldered over the hole in the apparatus.

Examples of Automata Machines

There are about 300 machines in the various treatises, so it would clearly be impossible to make a fully representative selection from these, even if they were described very briefly. Only a few examples are given, therefore, selected partly for their intrinsic interest, and partly because they incorporate features mentioned in the foregoing sections of this chapter. These features will be further illustrated in the discussion of water-clocks in the next chapter.

Tree, Birds and Snake Automaton — Philo [31]

Although this device is quite simple, it is included because the motif is common in Hellenistic works. It is unusual in that the wings of the bird have a movement of their own. It is typical of Philo that only general instructions for manufacture are given, and we are not told what materials the various parts are to be made of, except for the snake, which is made of silver. The tree (a), with the branches, nest and chicks has a hollow stem which is soldered to the perforated lid of the vessel (Figure 11.2a). A pipe (b) runs through the tree, and is soldered at the bottom to float (c), which, says Philo 'is like the float of a water-clock'. The hen bird rests on the top of pipe (b). Supports (ee) prevent float (c) from sinking to the bottom of the vessel. Rod (f) is soldered to the bottom of the vessel, goes through float (c) and pipe (b), and is connected to the bird's wings as shown in Figure 11.2b. There is a hole in the top of the bird, so that when pipe (b) pushes her upwards, the sides of her body thrust the wings outwards. The snake

Figure 11.2: Automaton from Philo

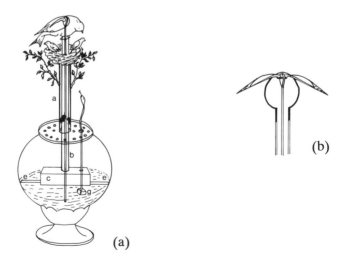

(b)

(a)

passes through one of the holes in the lid, through a hole in float (c) and is soldered to a small float (g).

When water or wine is poured in, the snake rises to menace the chicks; when the level rises far enough to lift float (c) the bird rises and spreads her wings. When the level falls (we are not told how this is made to happen) the bird folds her wings and settles down. In drawing Figure 11.2a it was necessary to make two emendations to Carra de Vaux's reconstruction; he omitted pipe (b) and showed the stem of the tree connected to float (c) — pipe (b) is mentioned in the text and was necessary, for without it the tree itself would rise. He also omitted supports (ee). These are not mentioned in the text, but there is probably a lacuna at this point. Without these supports both floats would rise together.

Aeolipile — Hero [32]

This essentially simple device, shown in Figure 11.3, has been frequently discussed because it is the first known case in which steam power was used to set a machine in motion. It is most unlikely, however, that this small device had any effect upon the much later development of the steam engine. The metal sphere, partly filled with water, is suspended between two trunnions. When heat is applied to the sphere, steam emerges from the two opposing jets, causing the sphere to rotate.

Self-feeding Lamp with Self-extruding Wick — Banū Mūsà [33]

This lamp combines two functions, which are shown separately in two of the Banū Mūsà's other lamps, and in two of Hero's.[34] To the right of

Figure 11.3: Hero's Aeolipile

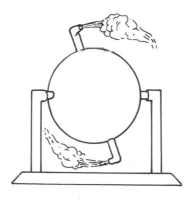

Figure 11.4: Lamp from Banū Mūsà

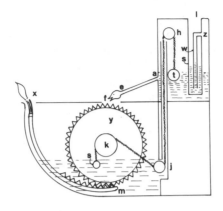

Figure 11.4 is a special filling pipe that prevents air from entering the sealed storage tank. In this tank is a float (t) to which a light chain is attached. The chain passes over pulley (h), goes down a pipe that opens into the body of the lamp, passes under a pulley, over pulley (k) and is attached to weight (s). On the same axle as pulley (k) is a large gear-wheel (y), that meshes with rack (m), on the end of which is the wick (x). A pipe (ae), its end in the shape of a bird's head, leads out from the reservoir and terminates above a hole in the cover of the lamp. Oil is poured in at (l) and runs into the tank through hole (z), and from the tank through pipe (ae) into the lamp until the hole at (j) is covered, whereupon flow ceases because no air can enter the tank. The wick is lit and as the oil in the lamp is consumed, hole (j) is uncovered, and a little more oil flows into the lamp. As the level in the tank falls float (t) descends, turning gear-wheel (y) through the pulley system. This moves rack (m), and a short length of wick is extruded. This is an automatic closed-loop system and since the frequency of the on-off cycle is high, the oil in the lamp is replenished without any noticeable variation in its level.

Trick Vessel — Banū Mūsà [35]

This device has been selected because it incorporates several of the Banū Mūsà's most commonly used techniques. It is by no means the most complex of their devices; if we assigned a scale of complexity to all their work from 1 to 10, this one would rate about 8. Figure 11.5 is traced from the illustration in one of the manuscripts (Topkapi Seray Mukesi A 3474) with Roman letters substituted for Arabic ones. The

Figure 11.5: Trick Vessel from Banū Mūsà

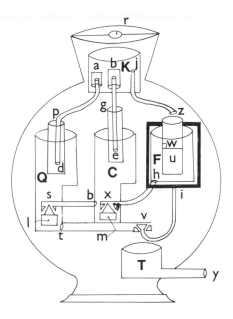

action of the device is as follows: three liquids of different colours are poured into hole (r) in the cover of the jar in succession; shortly after all the third liquid has been poured in, the liquids discharge from outlet pipe (y) in the same succession. As is the case with Philo and Hero, the Banū Mūsà give few constructional details; dimensions are not given, nor are we told how the small tanks and other components inside the large jar were supported. Nevertheless, there is no doubt that these trick vessels were made, and experiments confirm that they would have worked.

A tank (K) was installed in the neck of the jar, and two double concentric siphons and a pipe passed through holes in its floor, to which they were soldered. Top (a) of siphon (apd) was lower than top (b) of siphon (bge) which was in turn lower than top (j) of pipe (jz). These conduits led into tanks (Q), (C) and (F). Pipe (hx) led from tank (F) to tank (C), and pipe (bs) from tank (C) to tank (Q). At (x) and (s) the seats of conical valves were soldered to the pipes; beneath these were their plugs, carried on floats (m) and (l). (Actually, the lower sections of the tanks were probably narrower than shown on the drawing — just a little wider than the floats, so that these could not move out of position.) In tank (F) was float (u) with a small tank on top of it, having a small hole in its side at the bottom. A rod was bent in a rectangle around tank (F) and its ends were soldered to the sides of the small tank (w). At the centre of this rod at the bottom a valve-rod (iv) was soldered, bent upwards as shown and carrying the plug of an upward-opening conical valve (v), whose seat was in the end of pipe (tv) which led from tank (Q). Valve (v) was immediately above tank (T), to which the outlet pipe (y) was soldered.

The first liquid was poured in and ran through siphon (apd) into tank (Q). (Presumably the capacity of tank (Q) was known, so that exactly the right quantity was poured in — the same applies for the other two tanks.) Then the second liquid was poured in; because of the action of the double concentric siphon, the liquid could not flow through siphon (apd) so it flowed through siphon (bge) into tank (C). The third liquid flowed through pipe (jz) into the small tank (w), and overflowed into tank (F). The third liquid now dripped out of the hole in tank (w) into tank (F). When tank (w) was empty float (u) was able to rise, thus opening valve (v), and tank (Q) discharged its contents through pipe (tv) into tank (T) and out of the jar. Float (l) dropped,

Figure 11.6: Phlebotomy Measuring Device from al-Jazarī

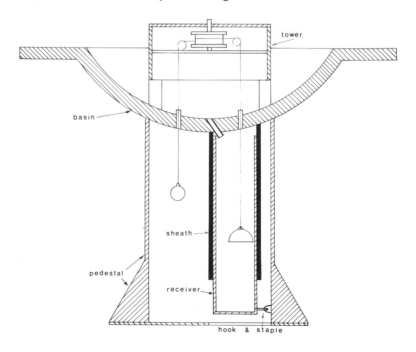

valve (s) opened, and tank (C) emptied by the route b-s-t-v-T-y. Finally, valve (x) opened, and the contents of tank (F) were discharged by the route h-x-b-s-t-v-T-y.

This device, and others like it, demonstrate an amazing skill in the use of differential pressures, and in the use of in-line valves for automatic control. Nothing like it is known to have been attempted before or since, until the advent of modern pneumatic instrumentation. Indeed, they had exhausted the subject, and it would have been impossible to emulate them in this kind of construction.

Phlebotomy Measuring Device — al-Jazarī [36]

This is the first of four phlebotomy measuring instruments described by al-Jazarī, and it is also the simplest, although all of them work on the same principle. It is called 'The Basin of the Monk'. Figure 11.6 has been drawn to the proportions determined by al-Jazarī's dimensions. The basin, made from brass, is about 18 inches in diameter, and it is supported on a hollow brass pedestal about 9 inches high at the centre. A copper cylinder 'the sheath' — of uniform cross-section is soldered vertically to the underside of the basin, a certain distance off-centre. Its lower end is open. A hole is made in the centre of the basin and a slanting pipe, terminating in the top of the sheath, is soldered to this hole. A second copper cylinder — 'the quiver' — is then made, also of uniform cross-section, its lower end closed. The quiver is then graduated: a measuring-cup of 5 *dirhams* (about 15 grammes) capacity is filled with blood or milk and poured into the quiver, and a mark is made coincident with the level of the liquid. This is repeated 24 times for a total of 120 dirhams, i.e. about 360 gr. The height from the bottom of the quiver to the '120 dirham' mark is recorded. Above the basin, supported on columns soldered to the floor of the basin, is a brass cylinder — 'the tower' — covered by a brass plate with a hole in its centre. Inside the tower a cross-beam is soldered diametrically. A large pulley is now made, its circumference equal to the length of the scale in the quiver. The pulley is installed in the centre of the tower, the lower end of its axle resting in a bearing on the cross-beam. The upper end of the axle passes through the hole in the cover of the tower — the section above the cover is square in cross-section. Two narrow pipes are soldered to holes in the basin, one above the quiver, the other diametrically opposite at the same distance from the centre of the basin. A small float is placed in the sheath, and a string is tied to a ring soldered to the top of the float, passed through the small pipe, over a small pulley, taken once around the large pulley, over a second small pulley, through the second small pipe, and is tied to a ring in the top of a lead weight. The rim of the basin is engraved with a circular scale divided into 120 divisions. The figure of a monk is made from sheet copper. A square hole in his foot fits over the top of the axle, and he carries a staff in his hand, the tip of which almost touches the divisions of the scale. The quiver is now inserted in the sheath and secured at the required level by a hook-and-staple fastening.

Before the blood-letting began, the basin was moistened with a little water. The blood was then conducted into the basin and ran into the quiver, causing the float to rise. The monk turned and the tip of his staff moved over the divisions of the scale. Extraction ended when the required quantity was indicated. The quiver was then removed and washed. (The method used here, whereby a figure was made to rotate by means of a float and pulleys, was used by al-Jazarī in a number of his clocks and automata machines.) In the other three devices of this kind, there were also displays that could be seen by an observer watching the face of the instrument, presumably so that patients of a nervous disposition could see how much of their blood was being extracted without rising from their seats. In the second device, for example, there were two seated scribes, one of whom, with a long pen, served exactly the same purpose. There was also a light rod soldered to the top of the float. This passed through a hole in the basin, through the hollow of one of the columns and through a hole in the tower's cover. Its upper, visible part was flattened to form a ruler, graduated into 120 divisions. This ruler passed through the hands of the second scribe, whose left-hand index finger pointed to the division of the scale corresponding to the quantity of blood extracted up to that moment. A full-size replica of this device, constructed for the 1976 World of Islam Festival, is now in the Science Museum in London.

Alternating Fountain — al-Jazarī [37]

All al-Jazarī's fountains change shape at regular intervals — an hour, half an hour or 15 minutes. In the introduction to this category he criticises the Banū Mūsà's fountains because they changed shape at irregular intervals, unpredictably. This is true, but it could be argued that this very unpredictability added to their attraction.

Figure 11.7 shows two adjoining tanks divided by a wall. On top of the wall was a fulcrum that supported a balanced pipe on top of which, in the centre, was a funnel into which the water was directed.

Figure 11.7: Fountain from al-Jazarī

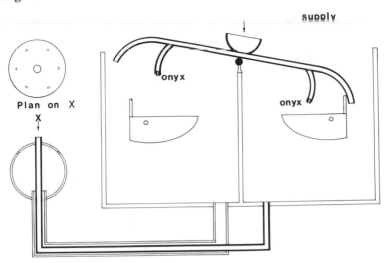

The ends of the pipe were open, and close to each end was a small pipe with a graduated orifice in its end. A tipping-bucket, suspended in each tank, was positioned below the orifice. In the situation shown in Figure 11.7, the pipe is discharging into the right-hand tank and the water is flowing through the narrow pipe and emerging from the fountainhead as a single jet. Meanwhile the tipping-bucket is slowly being filled from the orifice. When full it tilts, the projection on its end pushes the pipe, and this swings over to discharge into the other tank. The water flows through the wide pipe and emerges as six curved jets. The cycle repeats itself indefinitely as long as the water supply is maintained. The other fountains work in a similar manner: in some, floats with vertical rods soldered to their tops serve the same purpose as the tipping-buckets. There are also fountains with a double action, having two sets of pipes and two fountainheads.

Notes

1. Derek de Solla Price, *Science since Babylon* (Yale University Press, New Haven and London, enlarged edn, 1976), p. 50.

2. Ibid., p. 51.

3. Vitruvius, *De Architectura*, ed. F. Granger (2 vols., Loeb Classics, London, 1970), vol. 2, bk 9, Ch. 8, pp. 257-8.

4. Ibid., vol. 2, bk 7. preface, p. 75.

5. E. Wiedemann, *Aufsätze, sur arabischen Wissenschaftsgeschichte* (2 vols., Olms, Hildesheim, 1970), vol. 1, pp. 71-4.

6. Philo of Byzantium, 'Le Livre des Appareils Pneumatiques et des Machines Hydrauliques par Philon de Byzance', Arabic text ed. with French trans. by Carra de Vaux in *Paris Académie des Inscriptions et Belles Lettres*, 38 (1903), pt. 1.

7. *The Pneumatics of Hero of Alexandria*, ed. B. Woodcroft (Taylor, Walton and Maberly, London, 1851); facsimile edn with introduction by Marie Boas Hall (Macdonald, London and New York, 1971).

8. Donald R. Hill, 'Kustā b. Lūkā', in EI, vol. 5, p. 529.

9. Wiedemann, *Aufsätze*, vol. 1, pp. 69-70.

10. J. Ward Perkins, 'Nero's Golden House', *Antiquity*, 30 (1956) pp. 209-19.

11. The Banū Mūsà, *The Book of Ingenious Devices*, trans. and annotated by Donald R. Hill (Reidel, Dordrecht, 1979), pp. 3-6.

12. Ibid., p. 17.

13. Wiedemann, *Aufsätze*, vol. 1, pp. 188-228.

14. Donald R. Hill, *Arabic Water-clocks* (Institute for the History of Arabic Science, Aleppo, 1981) pp. 38-46.

15. Al-Jazarī, *The Book of Knowledge of Ingenious Mechanical Devices*, trans. and annotated by Donald R. Hill (Reidel, Dordrecht, 1974).

16. Al-Qalqashandī, *Subh al-a 'sha*, ed. Muhammad 'Abd al-Rasūl Ibrāhīm (14 vols., Cairo 1913-30), vol. 1, p. 477.

17. A.A. Dūrī, 'Baghdād' in EI, vol. 1, pp. 895-908, p. 898.

18. Gerard Brett, 'The Automata in the Byzantine "Throne of Solomon"', *Speculum*, 29 (1954), no. 3, pp. 477-87.

19. Al-Muqaddasī, *Ahsān al-taqāsīm*, ed. M.J. de Goeje, vol. 3 of BGA (Brill, Leiden, 1906), p. 157.

20. Ibn Jubayr, *Rihla*, ed. M.J. de Goeje (2nd. edn, Leiden, 1907), pp. 269-70.

21. John Webster Spargo, *Virgil the Necromancer* (Harvard University Press, Cambridge, Mass., 1934), pp. 8-26, 122-7.

22. Lynn White Jr, *Medieval Technology and Social Change* (Oxford University Press Paperbacks, 1964), p. 91.

23. *The Album of Villard de Honnecourt*, ed. H.R. Hahnlöser (Schroll, Vienna, 1935), Table 44.

24. Price, *Science since Babylon*, p. 65.

25. Ibid., pp. 65-6.

26. The Banū Mūsà, *The Book of Ingenious Devices*, pp. 26-7.

27. Philo, *Pneumatics* (Carra de Vaux), pp. 209-12.

28. Al-Jazarī, *The Book of Knowledge*, pp. 32-3.

29. The Banū Mūsà, *The Book of Ingenious Devices*, p. 32.

30. Joseph Needham, *Science and Civilisation in China* (5 vols. to date, Cambridge University Press, 1954 onwards) vol. 4, pt. 2 (1965), p. 537.

31. Philo, *Pneumatics* (Carra de Vaux), pp. 88-90 in Arabic, 176-8 in French.

32. Hero, *Pneumatics* (Woodcroft/Hall), p. 72.

33. The Banū Mūsà, *The Book of Ingenious Devices*, pp. 236-7.

34. Ibid., pp. 20-1.

35. Ibid., pp. 126-7.

36. Al-Jazarī, *The Book of Knowledge*, pp. 137-9, 260-1.

37. Ibid., pp. 157-8, 263.

12 Clocks

In the period from the beginning of the classical era to the end of the Middle Ages, about 2,000 years, the history of monumental water-clocks occupies about 1,600 years, that of the mechanical clock less than 200. Even if we extend the history of timekeeping to the present, the life of the mechanical clock is still less than half that of its predecessors. Nevertheless, almost all histories of horology, after devoting a few pages to sundials and water-clocks in their role as ancestors to the mechanical clock, then proceed to describe the invention of the verge escapement and the subsequent development of the mechanical clock.[1]

It is generally accepted that some of the features of water-clocks, astronomical instruments and automata came to be incorporated in mechanical clocks, but insufficient attention has been paid to such derivations, partly because information was lacking. A full description of the Antikythera Mechanism, for example, was not published until 1975, and the first mention of the complex gearing in al-Muradi's treatise two years later. Even so, the crucial features of the mechanical clock are the weight-drive and the escapement. Both of these are assumed to have been invented, almost without a pre-history, towards the end of the thirteenth century. It will be demonstrated, however, that the seeds of both ideas had been generating for some time before then.

In the preceding two chapters, simulation of celestial and biological behaviour was discussed, together with the mechanical, hydraulic and pneumatic means used to impart motion to these simulations. In water-clocks, as in mechanical clocks, these two types of representation were often combined, and an equal, or even greater, attention was paid to methods of producing, controlling and transmitting energy. Water-clocks are therefore of crucial importance in the history of mechanical engineering, and not simply for their contribution to the development of the mechanical clock. This is not meant to imply, however, that machines should be considered only in their relation to an evolutionary process, but also as satisfying creations in their own right. The mechanical clock is still an interesting, important (and often beautiful) machine, even though it is now being superseded by electronic timepieces.

Sundials, with a continuous history of at least 5,000 years, are of the greatest importance in the history of horology, but must nevertheless be excluded from this book because of their small engineering content. It is true that there are some moving parts in universal sundials, but the movements consist of raising and lowering plates against graduated arcs, opening concentric rings or moving cursors.[2] From our point of view, the main importance of the sundial lies in the fact that it cannot be used at night or on cloudy days; this disadvantage stimulated the development of water-clocks.

The Origins of Water-clocks

The earliest type of water-clock was the simple outflow clepsydra, an earthenware vessel with a hole in its side, near the base. It was in use in Egypt and Babylonia before 1500 BC and the oldest surviving example, from Egypt, is dated about 1380 BC. This had a scale of hours graduated upon its inside, which must have been drawn by observation so that allowance was automatically made for the decrease in the rate of discharge as the water-level fell. Attempts were made to minimise the effects of the steady decrease in the rate of discharge by tapering the profiles of the containers. The Oxyrhynchos Papyrus of the third century AD gives dimensions for the standard clepsydra: its profile was the frustrum of a cone having an upper radius of 12 fingers, a lower radius of 6 and a height of 18.[3] It is virtually impossible, however, to make an outflow clepsydra in which the water-level falls equal distances in equal successive time intervals. With very large vessels having large orifices the condition can be approached by making the profile a parabola in which the height is a given function of the radius. For vessels of a manageable size, however, this does not work, because with small heads of water and small orifices the effects of viscosity come into play, and a straightforward formula is no longer applicable.[4]

The outflow clepsydra was probably transmitted from the Fertile Crescent to other cultural areas during the first millenium BC. In Greek and Roman times it was used in the courts for allocating periods of time to speakers. In important cases, for instance when a man's life was at stake, it was filled, whereas for minor cases it was only partially filled. If the proceedings were interrupted for any reason, such as the examination of documents, the hole in the clepsydra was stopped with wax until the speaker was able to resume his pleading. The device was also used in Rome for timing each course (*missus*) of the Great Games in the Circus Maximus.

An important development was the introduction of the inflow clepsydra, which probably came into use in Egypt a little later than the outflow type. At first this consisted of two vessels, an outflow clepsydra that discharged into a receiver having an hour-scale marked on its interior. Clearly, this was no more accurate than the simple outflow type, but the next innovation made it a much better timekeeper. The reservoir vessel was provided with an overflow pipe and a constant supply of water, flowing in at a greater rate than the outflow from the orifice. A constant head was therefore maintained in the reservoir. The water supply could be from a large vessel, a stream or the city mains. The inflow clepsydra was an important component in some of the sophisticated water-clocks of Hellenistic times and later. Before discussing these, however, we must turn our attention briefly to the East, to establish a frame for cross-reference.

Water-clocks in India and China

In India mechanical devices, including war machines, were known as *yantras*; the word has roughly the same connotation as the Arabic *hiyal.* Unfortunately, Indian writers are more concerned with the effects produced by yantras than with the details of their construction (they are not alone in this). Furthermore, some of their descriptions are so fantastic as to be incredible. As a result, we know very little about mechanical technology in India. Nevertheless, there can be little doubt that there was a long tradition in India for the construction of yantras — descriptions occur in Indian works as early as 300 BC — and these included timepieces. Prince Bhoja, writing about AD 1050, mentions two types of timepieces. One of these had 30 figures lying prone around the circumference of a large open vessel: the whole thing revolved and in the centre was the figure of a lady who awoke one figure at the end of each time interval. The second was the figure of a rider on a chariot, an elephant or some other means of transport. For a fixed period the rider on his mount rotated, and at the end of the period a bell was sounded.[5]

Thanks largely to Joseph Needham, we are now well-informed about the use of timepieces in China. The inflow clepsydra was used extensively in China from the Han period (202 BC to AD 221) onwards. Two main types were developed. The first type involved the introduction of one or more compensating tanks between reservoir and receiver. At each successive stage the retardation of flow due to diminishing pressure head was more and more fully compensated. The introduction of at least one compensating tank can be dated to the second century AD. As many as six tanks are known to have been used. In the second type an overflow or constant head tank was placed in the series, a practice that began in the middle of the sixth century AD. Needham illustrates a variety of inflow clepsydras of both types, and combinations of the two, some of which have siphons in the upper vessels. There can be no doubt that this type of water-clock was widely used in China throughout the medieval period.[6]

The most notable achievement of Chinese horologists was the clock described by Su Sung in a book completed in AD 1090. This was a monumental clock, 30 to 40 feet in height, having a rotating armillary sphere and celestial globe, together with numerous jack-work figures with both audible and visible effects. These devices were driven, through a complex system of gearing, by a water-wheel 11 feet in diameter carrying 36 scoops on its perimeter. Water stored in an upper reservoir was delivered to a constant-level tank by a siphon, whence it was discharged on to the scoops of the wheel, each of which had a capacity of 0.2 cubic feet. The wheel was provided with a very ingenious escapement system, in essence two steelyards upon each of which the scoops acted in turn. When a scoop was full, its weight overcame the balancing system, and it fell freely for a given distance until checked, without recoil, by a locking device. The next scoop then came under the delivery jet and the cycle repeated itself.[7]

Steelyard clepsydras of a simpler type probably provided the inspiration for the escapement system of Su Sung's clock. These are first mentioned in a Chinese work of AD 450, with the implication that they had been in use some time before this date.[8] Between the sixth and fourteenth centuries AD they were used extensively in China. There were set rules for the operating crew, who were required to report the elapse of each major time interval to the responsible officials. There was a 'stop-watch' type of light clepsydra, but the more important design was a large balance having the receiver of an inflow clepsydra on the end of its shorter arm, and large and small weights on its longer arm. The weights were moved along the arm to balance the increasing weight of water in the receiver and so recorded the major and minor time intervals. Chinese historians mention a golden steelyard with 12 golden balls suspended from it that existed in Antioch in the seventh century AD.[9] It is apparent that the use of this device was widespread in the Old World before the advent of Islam, but the date of its origin and the direction of its transmission cannot yet be ascertained.

Hellenistic Water-clocks

Our earliest and most complete information about the water-clocks of the Hellenistic period comes from Vitruvius, in Chapter 8 (vol. 2, book 9) of *De Architectura*.[10] A very thorough and detailed analysis of this chapter was carried out by Drachmann, including a critical examination of the works of earlier European writers on the same subject.[11] Vitruvius attributes the invention of the inflow clepsydra, with overflow, and a float in the receiver with a vertical rod soldered to its top to Ctesibius. Vitruvius then goes on to describe how Ctesibius provided the rod with teeth that meshed with a toothed drum:

> In this, i.e. the float a rod is placed beside a drum that can turn. They are provided with equally spaced teeth, which teeth impinging on one another cause suitable turnings and movements. Also other rods and other drums toothed in the same way driven by one single movement cause by their turning effects and varieties of movements, by which statues are moved, obelisks are turned, pebbles or eggs are thrown, trumpets sound, and other by-works.

Vitruvius next describes how Ctesibius made another clock, also with a float and vertical rod. On top of the rod was a figure carrying a pointer which moved over a scale graduated into hours. He then wished to adapt the clock for the unequal hours, and he first attempted to do this by varying the flow from day to day by means of a valve. In his reconstruction based upon this passage, Diels has assumed that this was an automatically operating conical valve, but this is probably incorrect.[12] As Drachmann points out, the flow was to be varied by using wedges to hold the plug of the valve to its seat, and adjusting the wedges daily to give a greater or lesser aperture. Not surprisingly, this did not give accurate results. Vitruvius says that the wedges were often faulty, but the real fault lay in the method, since it would have been impossible to have obtained the very fine adjustments required by using wedges. Ctesibius therefore used another method. Instead of the linear scale, he erected a vertical drum, upon which were drawn lines for the months and the unequal hours. The drum was turned each day so that the pointer travelled over a distance representing the hours of daylight for that day (Figure 12.1).

Vitruvius describes two other clocks, but does not attribute them to Ctesibius. The first of these he calls the anaphoric clock, so called because the constellations of the Zodiac rose in succession over the horizon. Its motive power is an inflow clepsydra in which a float rises at constant speed. A soft bronze chain attached to the float is passed round a horizontal axle and has a sandbag at the other end as counterweight; the axle makes a complete rotation in 24 hours. Fixed to the axle is a disc upon which is a map of the stars of the northern hemisphere, with the North Pole in the centre. The ecliptic is shown as a circle eccentric in relation to the pole. In this circle 365 holes are drilled to take a small image of the sun, which is moved daily and acts as a pointer. The scale consists of a network of bronze wires placed in front of the disc. There are seven circles concentric with the rotating disc: the innermost circle is for Cancer, then there are circles for

Figure 12.1: Ctesibius' Water-clock

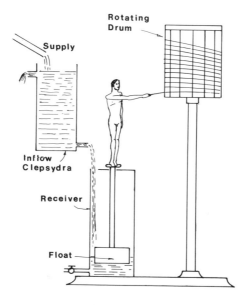

Gemini and Leo, Taurus and Virgo, Aries and Libra, Pisces and Scorpio, Aquarius and Sagittarius, and finally a single circle for Capricorn. Wires for the unequal hours run across these circles. When the sun is set to its correct position in the moving disc it moves in a circular arc through the unequal hours for that day. The rotatable disc and the network were probably drawn by stereographic projection. The clock is clearly a direct ancestor of the astrolabe. The fragment of a plate, dated to about the second century AD, was discovered near Salzburg and had been identified as part of the rotatable plate of an anaphoric clock, evidence that this device continued in use in the Roman Empire.[13]

The other description is concerned not with a complete clock but with a device for changing the outflow rate from the upper vessel into the receiver from day to day, to make allowance for the varying length of the unequal hours. The outflow from the reservoir was through a hole placed near the outer rim of a bronze disc which could be rotated. The names of the zodiacal signs were marked on the fixed annulus surrounding the disc, Cancer at the top, Capricorn at the bottom, Aries at the left and Libra at the right. The other signs were spaced between these, each sign occupying 30 degrees. This device embodies two incorrect ideas: that the waxing and waning of the days are symmetrically grouped around the equinoxes, and that the rate of flow is directly proportional to the head of the water.

Figure 12.2: Water-machinery from 'Archimedes' Clocks

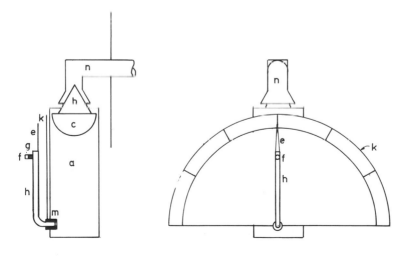

The earliest detailed description we have of a large water-clock is in a treatise attributed to Archimedes. This exists only in three Arabic manuscripts, all of which contain material other than the 'Archimedes' treatise.[14] The treatise is an important document in the history of water-clocks; both Ridwān and al-Jazarī acknowledge that the water-machinery in their monumental clocks was based upon the system described in the treatise, and there is also evidence (see below) that clocks based upon this design were built in the Persian and Byzantine Empires in pre-Islamic times. The water-machinery is shown in Figure 12.2. The float-chamber (a) is attached to the side of the clock, just below the reservoir. Plug (b) of a conical valve is soldered to the top of the small float (c) which rests on the surface of the water in the float-chamber. The plug enters the conical seat at the end of the pipe (n), which is the discharge from the reservoir. When water leaves the float-chamber through pipe (h) it rises and discharges through orifice (f) set in small pipe (g). The valve then opens momentarily, water flows into the float-chamber, whereupon the valve closes momentarily. There is therefore a closed-loop system operated by feed-back control, and the head in the float-chamber varies only very slightly. This is indeed a brilliant concept, but the flow regulator, for varying the head above the orifice from day to day for the unequal hours, is much less impressive. Pipe (h) can be rotated in bush (m), into which it is ground to a watertight fit. Pointer (e) attached to pipe (h) allows the orifice to be set to a position on the semicircular disc (k), which is divided into the twelve zodiacal signs. This is, of course, even more inaccurate than the full circle described by Vitruvius.

The clock consisted of four cylindrical vessels, erected vertically, and connected to each other by spigot-and-socket joints. The topmost vessel housed various mechanisms, the next was the reservoir which contained water upon which a large float rested — as the water discharged through the float-chamber the level in the reservoir fell at constant speed. Below the reservoir was a vessel containing mechanical and hydraulic components, and below this was an air-vessel. At the end of every unequal hour the eyes of a Gorgon at the top of the clock changed colour, and the head of a bird projecting from the upper vessel discharged a ball from its beak on to a cymbal. On the front of the vessel below the reservoir was a tree with a number of birds perched on its branches; two snakes emerged at the foot of the tree and the birds whistled. Finally, a flute-player in front of the bottom vessel played for a while.

A cord attached to the top of the float passed through concentric holes in the cover of the reservoir and the bottom of the upper vessel and was attached to the centre of a rotatable drum. On one end of the drum was a gear-wheel that meshed at right-angles with a horizontal lantern pinion. The axle of the lantern pinion passed through a fixed disc mounted on a cross-beam, and a rotating disc was fixed to its upper end. The lower disc had a single hole in it to which a pipe was attached that led to the head of the bird. There were twelve holes in the upper disc; at the start of the day there was a ball in each hole; at each hour one of the upper holes coincided with the hole in the lower disc, and a ball ran down the pipe and was ejected from the hinged beak of the bird on to the cymbal. Cords attached to the drum operated the mechanisms in the upper vessel through pulley systems. The birds-and-snakes and the flute-player were activated by the discharge of water from the orifice.

A number of other automata are described in the treatise, but these would have required to be housed in a different shape of container, since the movements they describe are horizontal. It is probable that these were later, Islamic additions. The earliest mention we have of this treatise is in the *Fihrist* of Ibn al-Nadīm, composed in AD 987/8. He refers to it as 'the book of a water-clock which discharges balls, by Archimedes'.[15] Later Arabic writers also attribute the basic design to Archimedes, and there seems no good reason for doubting this attribution. The Gorgon's head is, of course, a typically Greek motif. The snakes-and-birds automaton was, as we have seen, used by Philo, and indeed the mechanisms used to activate this device are also typical of Philo. It is therefore probable that the basic design of the clock was the invention of Archimedes, that additions were made by Philo and further additions by Islamic craftsmen.

This hypothesis is supported by the comments made by Ridwān in the introduction to a treatise written in AD 1203, describing the reconstruction of a clock built by his father in Damascus.[16] He describes the 'Archimedes' clock briefly but accurately. He does not elaborate upon the description, because, he says, the clock is very well known and 'we shall mention what he invented'. This included the float-chamber and its float; the reservoir and its float; one cymbal; the head of one bird; and one ball falling every hour. Ridwān says that this design was transmitted to Iran, where the geared transmission was changed to transmission by driving wheels and pulleys. Twelve doors, one of which opened every hour, were introduced, as was the head of a second bird. This design was transmitted back to the

Hellenistic world, and the tradition for constructing clocks of this type continued in Byzantine times, specifically in 'Damascus, where it was constructed up to the days of the Byzantines and after that in the days of the Banu Umayya, according to what is mentioned in the histories'. The construction of monumental water-clocks in the Byzantine Empire is confirmed by a description by Procopius of such a clock, built by an unknown craftsman in Gaza early in the sixth century.[17] No constructional details are given, but the twelve doors, mentioned by Ridwān and later to be a feature of Islamic clocks, are incorporated in this clock. Every hour an eagle over a door unfolded its wings and moved forward. The door opened, and the figure of Hercules emerged, carrying the spoils of one of his labours. There were other automata, all from Greek mythology, including a Gorgon, whose eyes rolled fearsomely every hour.

Islamic Clocks

According to Ridwān's statement, the tradition of constructing water-clocks in Syria was not broken by the advent of Islam. Writing late in the tenth century, al-Muqaddasī refers to the 'Gate of the Hours' in a portico by the Jayrūn Gate in Damascus, the place where Ridwān's father was to build his monumental clock in the twelfth century.[18] We can infer from al-Muqaddasī's remark that by the time he wrote it had become traditional to have a public clock by the Jayrun Gate. In the ninth century, an elaborate water-clock was presented to Charlemagne by vassals of Harūn al-Rashīd, an indication that these devices were made in early Abbasid times. In his *Book of the Balance of Wisdom*, written about AD 1121, al-Khāzinī refers to various water-clocks built by his predecessors, in particular the great Egyptian scientist Ibn al-Haytham (965-1039).[19] About the year 1050, the astronomer al-Zarqālī constructed a large water-clock on the banks of the Tagus at Toledo in Spain. This not only told the hours of the day and the night, but also recorded the phases of the moon. It consisted of two large basins, in a building, fed by underground conduits. As soon as the new moon appeared the basins began to fill, at the rate of one fourteenth of their capacity every day. When the moon began to wane, the basins emptied at the same rate, until on the twenty-ninth day both were empty. There was some kind of automatic compensating mechanism, so that if any water were extracted from or poured into one of the basins, the water immediately returned to its correct level. This clock was still in operation when the Christians occupied Toledo in 1085, and one of the clepsydras continued in service after 1133.[20]

Eighteen water-clocks are described in al-Murādī's treatise, composed in Andalusia in the eleventh century. The automata are of the usual kind: doors that open to reveal figures, figures that discharge pellets from their mouths, figures that rotate, etc. Model 8 has a set of mirrors of fine, white glass, one of which is illuminated every hour. The water-machinery is fairly crude — simply large clepsydras with long concentric siphons discharging from their undersides. Since the clocks recorded the passage of the unequal hours, however, some further refinement was necessary. The clepsydra was filled with the amount of water that would discharge in the 15 hours of daylight of the summer solstice at the latitude of Cordoba, a mark was made at water-level and the clepsydra was emptied. Another mark was made towards the bottom of the vessel, leaving an unmarked section at the bottom for the float. The clepsydra was then refilled, and the scale was graduated into the unequal hours and their divisions as the water descended. The components of the transmission system were dimensioned so that the time recording automata were activated at the end of each unequal hour. The quantities that would discharge in the daylight hours on other days in the year were known, so on a given day the required quantity was poured in, and the water-level was brought up to the top of the scale by placing solid objects in the vessel. This method was neither as accurate nor as convenient as using a float-chamber with a properly calibrated flow regulator. It is surprising that al-Murādī did not use this system, since he does use the 'Archimedes' geared method for releasing balls.

In the eighth treatise of his *Book of the Balance of Wisdom*, al-Khāzinī describes the construction of two steelyard clepsydras, a light one for 1-hour operation and a large one for 24-hour operation.[21] Although these machines were well constructed, with careful attention to detail, there is nothing very remarkable about their construction. What is worthy of attention, however, is al-Khāzinī's awareness of the physical properties of fluids. The clepsydras were designed to work either with water or with sand. In the large device a clepsydra is attached to the end of the short arm of the steelyard; a large weight and a small weight are suspended from the long, graduated arm. As the water or sand discharges, the weights are moved along the arm to bring the machine into balance. The hours are read from the position of the large weight, the minutes from the position of the small one. Al-Khāzinī gives careful instruction for the preparation of the materials. The water was to be clean and free from impurities and — since al-Khāzinī knew that its viscosity varied with temperature — the clepsydra was to be kept in a room of constant temperature. He also knew, of course, that the discharge slowed as the water-level descended, and his explanation for this anticipates the methods of integral calculus. Sand was to be washed twice to free it from dirt and dust, passed through two sieves of different mesh sizes, then stored in a covered container. Care was to be taken to avoid air inclusions.

The first description we have of a large monumental water-clock in Islam occurs in a treatise written by Ridwān b. al-Saʿātī (son of the clockmaker) in AD 1203. This describes the reconstruction of the clock built by his father Muhammad in the reign of Nūr al-Dīn Mahmūd b. Zenkī, who resided in Damascus from 1154 until his death in 1174. The clock, built at the Jayrūn gate in Damascus, fell into disrepair after Muhammad's death, and his son undertook its repair after others had failed in the task. His treatise does not simply describe the repairs, but gives full constructional details of the clock.

The front face of the clock, as seen by observers, is shown in Figure 12.3. A wall of well-seasoned timber was installed on a masonry lower wall and between masonry side walls, the whole forming the front wall of a chamber about 4.83 metres square by 4.88 metres in

Figure 12.3: Face of Ridwān's Clock

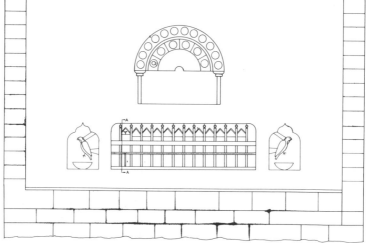

height (in Figure 12.3 the lower part of the front masonry wall has been omitted). The wooden wall was about 4.23 metres wide by 2.78 metres high. The lower part of the clock was for the 'hours of the day'. There was a row of twelve doors above each of which was a shutter. A beam studded with gilded nails ran in front of the doors and a vertical rod with a small crescent on its upper end moved slowly along in front of the beam. The nails represented the divisions of the unequal hours. At each end of the wooden screen was a niche containing the figure of a falcon. On the hour, when the crescent reached the centre of the first door, the door rotated on its vertical axle and displayed its rear, which was painted a different colour from the front. The shutter came upright to reveal its longer arm, carrying a small crescent on its top. A ball ran into the beak of each falcon, which then tilted forward, opened its beak, and discharged the ball on to a cymbal in a bowl, causing a ringing sound. This happened at each hour, until at the end of the day all the doors had changed colour, all the shutters were erect and twelve balls had been discharged by each falcon.

The 'hours of the night' and the 'hours of the sun' are shown by the outer and inner semicircular bands above the doors. The circles within the upper band are holes cut through the main wall, and the semicircles are simply lines painted on the woodwork. The inner set of circles are holes cut around the outer edge of a large wheel that rotated outside the main wooden screen. Each hole was in one of twelve equal divisions, each marked with a sign of the Zodiac. In the appropriate hole for the time of the year was a small disc representing the sun. It was set to one of three positions in the sign, depending upon whether it was the beginning, middle or end of the zodiacal period. The wheel turned through 180 degrees in twelve unequal hours. Half of the wheel was concealed by a wooden screen fixed to the outside of the wall; this was the 'horizon' for the sun/zodiac wheel. As an hour of the night passed, one of the outer circles became illuminated, and was fully illuminated at the end of the hour. At the end of the night all the circles were illuminated. At daybreak the 'sun' rose in its correct sign and the wheel rotated, turning through 180 degrees by sunset. Every hour one sign appeared and another disappeared.

The water-machinery was similar to that of the 'Archimedes' clock except that the flow regulator was a full circle. Two pulley systems were operated from the float in the reservoir. One of these turned the sun/zodiac wheel and another wheel that uncovered the holes for the 'hours of the night'. Through the second system a slider was drawn along a metal track behind the wooden wall. It carried the crescent, which projected through a slit in the wall, and a vertical rod. At the end of each hour the rod activated the mechanisms that caused the doors and shutters to rotate, together with the ball-release device. The balls rolled down sloping channels, and entered the heads of the falcons, who tilted forward and ejected the balls on to the cymbals. Apart from the inaccuracy of the flow regulator, the main defect of this clock is its structural weakness. The long wooden wall was unsupported except at its ends, and the location of the reservoir in the lower masonry wall meant that all the mechanisms had to be supported by the wooden wall. The timber would have been liable to warping, not to mention damage by fire. This vulnerability is confirmed by its destruction by fire about 1167, and its falling into disrepair after it had been rebuilt by Ridwān's father.[22] This was presumably after 1184, since in that year the clock was seen and described by Ibn Jubayr.[23]

The first, and most elaborate, of the water-clocks described by al-Jazarī was similar in many ways to the one built by al-Sa'ātī, but there were some important differences. The working face was a screen of bronze or wood about 1.35 metres wide by 2.25 metres high, with another 0.75 metres for the semi-diameter of the sun/zodiac circle, which in this case was at the very top of the clock. (The 'hours of the night' were lower down the clock-face.) Again, there was a moving crescent and two falcons, who spread their wings when they leant forward to eject the balls on to the cymbals. The lower set of twelve doors rotated to show a different colour, as did the lower set in al-Sa'ātī's clock, and above these was a second row of twelve doors having double leaves which opened on the hour to reveal standing figures. In addition, on a platform in front of the clock, at the bottom, were the figures of five musicians — two trumpeters, two drummers and a cymbalist. Water dripping from the orifice accumulated in a special tank, from which it was suddenly released at the sixth, ninth and twelfth hours. It ran on to the scoops of a water-wheel, which rotated. On its axle were cams which bore down on projections from the arms of the percussion players. From a tank below the water-wheel the water ran into an air-vessel, to which a mechanical whistle was attached, for simulating the sound of the trumpeters.

The working face was set in the front wall of a wooden chamber, and the water-machinery was positioned inside this chamber, transmission to the automata being by means of axles. Very little load was carried by the working face, which in any case was narrower and more robust than al-Sa'ātī's. Al-Jazarī tested various types of flow regulator, including the full circle divided into equal divisions and the semicircle of 'Archimedes' and found that they were inaccurate. He tells us how he painstakingly graduated his circle by trial-and-error to produce an accurate instrument.[24] A full-scale replica of this clock, using al-Jazarī's instructions and techniques, was built in the Science Museum, London for the 1976 World of Islam Festival. It worked perfectly.

In his third and fourth clocks al-Jazarī used the submersible bowl (*tarjahār*) as the driving mechanism, the only water-clocks known to have used this device.[25] Figure 12.4 shows the basic principles; it would be impossible to describe either of the clocks fully. The fourth one in particular — The 'Elephant' clock — had a number of automata, and the mechanisms for activating them, some of them very ingenious, were neatly assembled in a confined space. The bowl (a) rests on the surface of the water in tank (n), to which it is connected by the three flat pin-jointed links (b). A rod is soldered across a diameter of the bowl, with hole (k) in its centre. At the top of the clock,

Figure 12.4: Water-machinery from al-Jazarī's Clocks

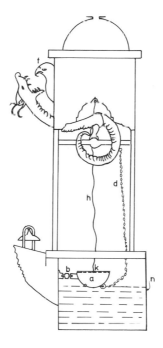

supported on four columns is the 'castle', a square brass box with a detachable dome. Inside the castle is a ball-release mechanism (not shown) from which a channel leads to the head (f) of a bird. The tail of the serpent, in effect a pulley, rotates on an axle that rests in bearings in transome fixed between each pair of columns. The open mouth of the serpent is just below the head of the bird. A light chain (d) runs from the underside of the bowl to a staple in the tail of the serpent. A wire (h) is tied to a hole (k) and to the ball-release. At the beginning of a time period — an hour or half an hour — the empty bowl is on the surface of the water. It sinks slowly, until at the end of the period it suddenly submerges. Wire (h) operates the ball-release, and a ball runs into the bird's and out of its hinged beak into the mouth of the serpent. The serpent's head sinks, and chain (d) lifts the bowl, which tilts due to the combined pull of the chain and links (b) and discharges all its contents. The ball drops from the serpent's mouth on to a cymbal, and the serpent's head rises to its previous position. The empty bowl is once again horizontal on the surface of the water, and the cycle re-starts. This is another closed-loop system — the clock will continue working as long as there are balls in the magazine.

The candle clock shown in Figure 12.5 is one of four described by al-Jazari.[26] The clock is designed to record the passage of 14 equal hours. The candle is to be of uniform cross-section and its weight and the weight of its wick are specified. A long brass sheath is soldered into a hollow brass candle-holder — the total height of the clock is about three feet. A cap for the sheath is made from cast bronze; it has a hole in its centre and an indented section around the hole on the top. The underside of the cap is machined so that it is truly flat and horizontal when fitted to the top of the sheath (by a bayonet fitting). A flat copper dish with a rim is made to fit the bottom of the candle. A long, flat iron rod is soldered to the centre of the underside of the dish; the rod has a hole in its lower end. Just below the dish, a light copper channel is made with a U-shaped cross-section; its upper part is divided into 14 compartments. It is suspended by a hook-and-staple connection to the dish under the candle, its open side towards the side of the sheath. It is kept in alignment by a guide channel soldered to the candle-holder. A U-shaped lead weight is now made; staples are fitted to the centre of each leg at the top, and a small ring at the bottom. The weight fits round the iron rod and the channels in a loose sliding fit, its open end towards the sheath. A string is tied to each of the staples on the side of the weight, passed over one of the small pulleys on the rod, and tied to the hole in the bottom of the rod. A

Figure 12.5: Candle Clock from al-Jazarī

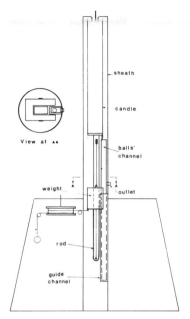

string, passing through a slit in the sheath, connects the ring at the bottom of the weight to a pulley system, like the one used to turn the figure of the monk in the phlebotomy measuring device. The figure of a scribe (not shown) sits on top of the protruding end of the axle — the tip of his pen almost touches a circular scale, graduated into hours and minutes. The outlet at the left enters the head of a bird (not shown).

The candle is fitted to the dish, the wick is drawn through the hole in its centre, and the cap is made secure. The wick is lit and the candle begins to burn away; the wax runs into the indentation in the top of the cap, which is cleaned out regularly to ensure that the wick burns with a steady flame. The candle is slowly forced up by the weight acting through the pulleys, and the scribe's pen moves round the scale. After an hour, the first of the 14 balls which have been loaded into the channel reaches the outlet. It rolls into the head of the bird and emerges from its hinged beak. The balls collect in an indentation in the cover of the holder. All four candle clocks are examples of small-scale precision engineering.

That the tradition for the building of monumental water-clocks continued in Islam at least until the fourteenth century is proved by the remains of two such clocks, dated to that time, in Fez, Morocco. These seem to have been very similar to al-Sa'ātī's clock, except that the working faces were built of masonry.[27] The clocks in the *Libros del Saber* are also of Islamic origin, since the work is a compilation of translations and paraphrases from Arabic sources. It is a little disappointing, however, to discover that the water-clock and the candle clock described in the *Libros* are less advanced than those of al-Jazarī. The water-clock is essentially an inflow clepsydra with pneumatic feed-back control. The water-level is kept constant in a small tank connected to the reservoir by a method similar to that used by the Banū Mūsà for the automatic feed to their lamp. It has sometimes been said that this clock had an astrolabic dial, but this is not the case. A vertical board fitted to the float in the receiver, and called the *semiante del cielo* — 'the likeness of the sky' — was marked into various divisions for equal and unequal hours, Roman time, the signs of the Zodiac and astronomical events. The board rose behind the bars of a 'ladder', and the time could be observed by the passage of the divisions behind the horizontal bars of the ladder. The candle clock was mostly made of wood and acted on the same principle to al-Jazarī's in that the candle was forced up by a weight as it burned away. Instead of a properly fitted cap, however, it was held down by two pieces of wood and would have suffered from a defect observed by al-Jazarī in some earlier candle clocks, namely that the wax ran down and clogged the working parts.[28]

The third clock is more interesting. It consisted of a large drum made of walnut or jujube wood tightly assembled and sealed with wax or resin. The interior of the drum was divided into twelve compartments, with small holes between the compartments through which mercury flowed. Enough mercury was enclosed to fill just half the compartments. The drum was mounted on the same axle as a large wheel powered by a weight-drive wound around the wheel. Also on the axle was a pinion with six teeth that meshed with 36 oaken teeth on the rim of an astrolabe dial. The mercury drum and pinion made a complete revolution every 4 hours and the astrolabe dial made a complete revolution in 24 hours.

This type of timepiece had been known in Islam since the eleventh century — at least 200 years before the first appearance of weight-driven clocks in the West. The *Libros del Saber* was translated into Italian in 1341, but it was not until 1598 that a similar clock was described in a treatise by Attilio Parisio. It seems likely, although it remains to be proved, that Parisio's clock was derived from the one in the *Libros*. A few years later, Solomon de Caus, a German specialist in water mechanisms (1576–c. 1635) described a compartmented cylinder clock that was to be the prototype of many examples constructed in exactly the same form which were to appear again and again in the seventeenth and eighteenth centuries.[29] Clearly, the compartmented drum, whether filled with mercury or with water, was an effective escapement for weight-driven clocks.

European Water-clocks

There is sufficient evidence to show that water-clocks equipped with some kind of alarm device were in use in western Europe during the Middle Ages. For example, there is a Latin text of the Cistercian rule, dating to the early part of the twelfth century, that prescribes rules for the sounding of the alarms for services. Part of Rule CXIV reads: 'The Sacrist: the Sacrist shall set the clock (*horologium temporare*) and cause it to sound (*facere sonare*) in winter before lauds ...'[30] About 80 years later we find similar instructions in Dante: 'One says *mezza Terza* before ringing for that part of the day, and *mezza Nona* after that part is rung; similarly *mezzo Vespro*' (*Convito* IV, Ch. 23). The verb translated here as 'ring' is *suonare*, and this could apply to the ringing of a clock or the manual striking of a bell. Dante was able to hear these signals during the night, because he describes how he awakened from a dream and then found that the hour in which the vision had appeared had been the fourth hour of the night.[31]

The earliest description we have of a European water-clock appears in Latin in MS Ripoll 225 from the Benedictine monastery of Santa Maria de Ripoll at the foot of the Pyrenees. The manuscript is now in the Archivo de la Corona de Aragon in Barcelona. Estimates for its date vary from the middle of the tenth century to some time in the eleventh. The description of the main water-machinery is missing but the section describing the striking train is complete. Although this is not very clearly described, it appears that cams on a wheel driven by the water-machinery released weights at intervals, and the release of the weights caused the oscillation of iron arms that struck bells.[32] An illustration in a manuscript dated about 1285 shows a water-clock in a monastery in northern France. It is not possible to determine the precise operation from the illustration, but it seems that there was a reservoir below which an inflow clepsydra was suspended. A chain or cord from this receiver passes round the axle of a wheel. The wheel is divided into 15 segments, and between each pair of segments is a hole and a projection on the perimeter. A row of bells above the wheel indicates that this was a chiming clock.[33] Sleeswyk has suggested that the bells were rung at intervals by mechanisms activated by the steady sinking of the inflow clepsydra.[34]

The existence of a considerable market for water-clocks in Europe in the twelfth century is implied by the establishment of a guild of clockmakers in Cologne in 1183.[35] The problem of the origin of medieval European water-clocks is unresolved. They may derive from early transmissions from Islamic sources, or may continue the late Roman tradition of the anaphoric clock. Given the tradition for constructing and writing about water-clocks in Islamic Spain, however, a diffusion from Islam seems likely. The stimulus could, of course, have come from both sources.

The Mechanical Clock

We have seen that many of the ideas that were to be embodied in the mechanical clock had been introduced centuries before its invention: complex gear trains in the Antikythera Mechanism, al-Bīrūnī's calendar and al-Murādī's treatise; segmental gears in al-Murādī and al-Jazarī; epicycle gears in al-Murādī; celestial and biological simulations in the automata-machines and water-clocks of Hellenistic and Islamic engineers; weight-drives in Islamic mercury clocks and pumps, and in the striking trains of European water-clocks; escapements in Su Sung's clock and in Islamic mercury clocks. The heavy floats in water-clocks may also be regarded as weights, with the constant-head system as the escapement. Al-Jazarī's remarks, in connection with controlling the speed of a water-wheel that was operating a musical automaton, are also relevant (he is referring to the drawing of an old machine): 'And I say that even if the wheel caused a number of rods to fall in succession it would not be slow enough to display the changes of shape ...'[36]

The foregoing should not be taken as implying that the makers of instruments, water-clocks and automata were working towards a predetermined but unseen objective, namely the invention of the mechanical clock. Their work bore fruit in many other fields but, in any case, we have no way of knowing how much of the array of ideas listed in the previous paragraph was in the mind of the inventor of the

Figure 12.6: Verge Escapement

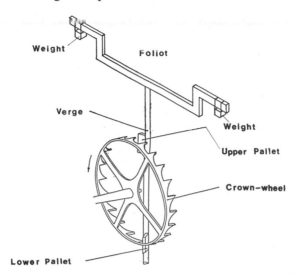

mechanical clock. Nevertheless, it is much more likely that the final invention was the result of a synthesis of earlier ideas, with one vital idea added, rather than a complete creation from the void. Indeed, in a treatise written in 1271, Robertus Anglicus tells us that the makers of horologia were trying to make a weight-driven clock. 'But they cannot quite complete their task which if they could, it would be a really accurate horologium.'[37] This is a clear indication that the invention of the escapement was approaching at that time, and also implies that the invention was made by a clock-maker who was already familiar with the construction of water-clocks.

The earliest surviving form of escapement has come to be known as the *verge escapement* and it was one of the fundamental inventions in the history of technology. Figure 12.6 shows the essentials of the system. The crown-wheel, which has an uneven number of teeth is connected by gears to the weight-driven drum of the clock; a shaft or 'verge' was provided with rectangular projections known as 'flags' or 'pallets', and on top of the shaft was either a horizontal bar called a 'foliot' or a large wheel called a 'balance'. As the weight is applied to the 'going-train' of wheels meshing one into the other, the crown-wheel is turned. The top pallet (say) is thereby pushed aside by a tooth on the moving crown-wheel and in consequence the shaft is turned and so is the foliot on top of the shaft. During this turn the tooth on the moving crown-wheel 'escapes' from contact with the top pallet; at the same moment the lower pallet is caught by a tooth on the lower part of the crown-wheel, which is of course moving in the opposite direction to the top of the wheel. Because there is an uneven number of teeth on the crown-wheel, when the lower pallet is engaged by a tooth, the upper pallet will coincide with a space between two teeth. The shaft or 'verge' is impelled to move in one direction by the force of the tooth of the crown-wheel on the pallet until it 'escapes', at which moment the other pallet becomes engaged by a tooth on the opposite side of the crown-wheel, so impelling the verge to move in the opposite direction. The speed of this turning motion backwards and forwards is adjustable, either by applying a different weight to the going-train of the clock, or by moving the tiny weights on the foliot further out on the arm (to make it slower), or closer in on the arm (to make it swing faster).

The striking mechanism is a quite separate element, though it is released by the going-train of the clock. A separate weight provided the force for this 'striking-train', which was set in motion when a pin on one of the wheels of the going-train, upon reaching the hour, lifted a lever. The number of times that the bell is struck is regulated by the 'count-wheel', a wheel with notches cut in it at increasingly wide distances from each other. While the lever is held up by the count-wheel, the bell goes on being struck by a hammer; the longer the lever is kept up by the increasingly wide space between each notch, the more times the bell is struck by the hammer. The verge escapement and the count-wheel mechanism were the only means for regulating clocks for over 300 years. The former was superseded by the invention of the pendulum clock in 1657, and the latter by the invention of the 'rack-striking' device in 1676.[38]

Almost certainly, the mechanical clock was invented in north-western Europe but its inventor and the precise date of its introduction are unknown. It is unlikely that we shall ever know the inventor's name, and it is extremely difficult to establish a probable date for the invention; the subject has been the subject of much debate among historians of horology. The problem is partly one of vocabulary. The word *horologium* can mean any kind of timepiece, and we can therefore never be certain, when we read of the construction of a large horologium, whether this was a water-clock or a mechanical clock, except in the rare cases where a description is given or when the clock in question still exists. And even when a clock has survived, we cannot always be sure that it has not been modified over the centuries. We can, however, be sure that the event occurred between 1271 when Robert Anglicus wrote his treatise and 1348. The Dover clock (now in the Science Museum, London) carries the date 1348, and this is also the year when Giovanni de' Dondi began to construct his astronomical clock, driven by weights with a verge escapement.[39] Between 1286 and 1340, however, a number of clocks were built in England, France and northern Italy and it is probable that several of these were mechanical clocks. It is in fact certain that the clock completed in Norwich Cathedral in 1325 and the one built by Richard of Wallingford in St Albans Abbey about 1327 were mechanical clocks. The escapement in Richard of Wallingford's clock, however, was not of the verge and crown-wheel type, although it was probably effective as a regulating mechanism.[40]

It seems likely that the construction of mechanical clocks began towards the end of the thirteenth century, and that two or three types of escapement were tried out before it became apparent that the verge type gave the most satisfactory results. After about the middle of the fourteenth century, the mechanical clock spread with astonishing rapidity throughout western Europe. To quote Lynn White:

> Suddenly, towards the middle of the fourteenth century, the mechanical clock seized the imagination of our ancestors. Something of the civic pride which earlier had expended itself in cathedral-building now was diverted to the construction of astronomical clocks of astounding intricacy and elaboration. No European community felt able to hold up its head unless in its midst the planets wheeled in cycles and epicycles, while angels trumpeted, cocks crew, and apostles, kings, and prophets marched and countermarched at the booming of the hours.[41]

The clocks of Giovanni de' Dondi and Richard of Wallingford incorporated very complex gearing, and all the medieval clocks contained intricate mechanisms for activating the automata. The mechanical clock can be seen as the summation of a tradition that began in Hellenistic times and continued in Islam. It is also, however, one of the main foundations for the development of machine technology in subsequent centuries.

Notes

1. For example, A.P. Usher, *A History of Mechanical Inventions*, 2nd edn (Harvard University Press, Cambridge, Mass., 1954), pp. 191ff., and 304ff.
2. A comprehensive work in sundials is: René R.J. Rohr, *Sundials*, trans. from the French by Gabriel Godin from the original 1965 edn (University of Toronto Press, Toronto, 1970).
3. L. von Borchardt, *Die Alatägyptische Zeitmessung* (Berlin and Leipzig 1920), vol. 1 of *Die Geschichte der Zeitmessung und die Uhren*, ed. E. von Bassermann-Jordan (3 vols. 1920-5), p. 11.
4. Donald R. Hill, *Arabic Water-clocks* (Institute for the History of Arabic Science, Aleppo University, 1981), pp. 4-5; a thorough analysis of the fluid mechanics of clepsydras is: A.A. Mills, 'Newton's Water Clocks and the Fluid Mechanics of Clepsydrae', *Notes and Records of the Royal Society of London*, vol. 37, no. 1 (1982), pp. 35-61.
5. V. Ragavan, 'Yantras or Mechanical Contrivances in Ancient India', *The Indian Institute of Culture*, Basavangudi, Bangalore, Transaction No. 10 (1952), p. 23.
6. Joseph Needham, *Science and Civilisation in China* (5 vols. to date, Cambridge University Press, 1954 onwards), vol. 3 (1959) pp. 320ff.
7. Ibid., vol. 3, pp. 359ff., and vol. 4 (1965), pp. 446-66.
8. Ibid., vol. 3, pp. 326ff.
9. Joseph Needham with Wang Ling and Derek de Solla Price, *Heavenly Clockwork* (Cambridge University Press, 1960), p. 88.
10. Vitruvius, *De Architectura*, ed. F. Granger (2 vols. Loeb Classics, London, 1970), vol. 2, pp. 255-67.
11. A.G. Drachmann, 'Ktesibios, Philon and Heron; a Study in Ancient Pneumatics', *Acta Historica Scientiarum Naturalium et Medicinalium*, ed. Biblioteca Universitatis Hauiensis, Copenhagen, 4 (1948), pp. 19-36.
12. H. Diels, *Antike Technik* (Leipzig and Berlin, 1914), pp. 205-7.
13. Ibid., p. 213ff.
14. Donald R. Hill, *On the Construction of Water-clocks*, Occasional Paper no. 4 (Turner and Devereux, London, 1976). *passim.*
15. Ibn al-Nadīm, *Kitāb al-Fihrist*, ed. G. Flugel (2 vols., Leipzig, 1871-2; reprinted in 1 vol., Cairo, 1929/30), Cairo edn, p. 372.
16. Hill, *Arabic Water-clocks*, pp. 12-13.
17. H. Diels, 'Über die von Prokop beschriebene Kunstuhr von Gaza', *Abhandl. d. preuss. Akad. Wiss. Berlin* (Phil. Hist. Klasse, 1917), no. 7, *passim.*
18. Al-Muqaddasī, *Ahsān al-taqāsīm*, ed. M.J. de Goeje, vol. 3 of BGA (Brill, Leiden, 1906), p. 158.
19. Hill, *Arabic Water-clocks*, p. 49.
20. J.M. Millás Vallicrosa, *Estudias sobre Azarquiel* (Madrid, 1950), pp. 6-9.
21. Hill, *Arabic Water-clocks*, pp. 47-68.
22. Ibid., pp. 69-88.
23. Ibn Jubayr, *Rihla*, ed. M.J. de Goeje (2nd edn, Leiden, 1907).
24. Al-Jazari, *The Book of Knowledge of Ingenious Mechanical Devices*, trans. and annotated by Donald R. Hill (Reidel, Dordrecht, 1974), pp. 17-41.
25. Ibid., pp. 51-70.
26. Ibid., pp. 83-93.
27. Derek de Solla Price, 'Mechanical Water Clocks in the 14th Century in Fez, Morocco', *Ithaca*, 26, VIII-2 IX (1962), Hermann, Paris, pp. 599-602.
28. M. Rico y Sinobas, *Libros del Saber de Astronomia* (5 vols. Madrid, 1863), vol. 4, bk. 2 — water-clock; bk. 3 — candle clock.

29. Silvio A. Bedini, 'The Compartmented Cylindrical Clepsydra', *Technology and Culture*, 3 (1962), pp. 115-41.

30. C R Drover, 'A Medieval Monastic Water clock' in Antiquarian Horology, 1 (1954), pp. 54-8 and 63. The passage is on p. 55.

31. *Vita Nuova*, III.

32. Francis Maddison, Bryan Scott and Alan Kent, 'An Early Medieval Water-clock', *Antiquarian Horology*, 3 (1962), pp. 348-53.

33. Drover, 'A Medieval Monastic Water-clock'.

34. Andre Sleeswyk, 'The 13th Century "King Hezekiah" Water-clock', University of Groningen, unpublished paper.

35. Lynn White Jr, *Medieval Technology and Social Change* (Oxford University Press paperback, Oxford, 1964), p. 120.

36. Al-Jazari, *The Book of Knowledge*, p. 170ff.

37. Lynn Thorndike, 'Invention of the Mechanical Clock about 1271 AD', *Speculum*, 16 (1941) pp. 242-3.

38. Hugh Tait, *Clocks in the British Museum* (Trustees of the British Museum, London, 1968), pp. 17-19.

39. H.A. Lloyd, *Giovanni de' Dondi's Horological Masterpiece* (Hookwood, Limpsfield, Oxted, Surrey, 1956).

40. *Richard of Wallingford*, ed. with English trans. and commentary by J.D. North (3 vols., Oxford University Press, Oxford, 1976), vol. 2, pp. 330-4.

41. Lynn White Jr, *Medieval Technology*, p. 124.

Bibliography

Armytage, W.H.G. *A Social History of Engineering*, 3rd edn (Faber and Faber, London, 1970)

Backinsell, W.G.C. *Medieval Windlasses*, Historical Monograph 7 (South Wiltshire Industrial Archaeological Society, Salisbury, 1970)

al-Balādhurī, Aḥmad b. Yaḥyà, *Futūh al-Buldān*, Arabic text with Latin critical apparatus, published as *Liber Expugnationis Regionum*, ed. M.J. de Goeje (Brill, Leiden, 1866)

Banū Mūsà, *The Book of Ingenious Devices*, trans. and annotated by Donald R. Hill (Reidel, Dordrecht, 1979)

_____ *Kitāb al-ḥiyal*, Arabic text of the above, ed. A.Y. Hassan (Institute for the History of Arabic Science, Aleppo University, 1981)

Bedini, Silvio A. 'The Compartmented Cylindrical Clepsydra', *Technology and Culture*, 3 (1963) pp. 115-41

Bedini, Silvio A. and Maddison, Francis R. 'Mechanical Universe, the Astrarium of Giovanni de' Dondi', *Transactions of the American Philosophical Society*, new series, vol. 56, 5 (1966)

al-Bīrūnī, 'Abd al-Rayhān, *Kitāb al-jamāhir fī ma'rifat al-jawāhir*, ed. F. Krenkow (Hyderabad, Deccan, 1936)

Bishop, M. *The Penguin Book of the Middle Ages* (Penguin Books, London, 1971)

von Borchardt, L. *Die Altägyptische Zeitmessung* (Berlin and Leipzig 1920), vol. 1 of *Die Geschichte der Zeitmessung und die Uhren* ed. E. Von Bassermann-Jordan (3 vols., de Gruyter, Berlin and Leipzeg 1920-5)

Boyer, Marie N. 'Water-Mills: A problem for the Bridges and Boats of Medieval France', *History of Technology*, vol. 7 (1982) pp. 1-22

Brett, G. 'The Automata in the Byzantine "Throne of Solomon"', *Speculum*, 29, no. 3 (1954) pp. 477-87

Briggs, M.S. 'Building-construction' in *A History of Technology*, ed. Singer *et al.*, vol. 2, pp. 397-448

Bromehead, C.N. 'Mining and Quarrying to the Seventeenth Century' in *A History of Technology*, ed. Singer *et al*, vol. 2, pp. 1-40

Burns, R.I. *Medieval Colonialism* (Princeton University Press, Princeton, 1975)

Cahen, C. 'Le Service de l'irrigation en Iraq au début du XIe siècle', *Bulletin d'études orientales*, vol. 13 (1949-51), pp. 117-43

Calvert, N.G. 'On Water Mills in Central Crete', *Transactions of the Newcomen Society*, vol. 45, pt. 2 (1972-3), pp. 217-22

Citarella, A.O. 'The Relations of Amalfi with the Arab World before the Crusades', *Speculum*, 42 (1967), pp. 299-312

Collett, J. 'Water Powered Lifting Devices', unpublished paper prepared for the Land and Water Development Division, FAO, 1980

Creswell, K.A.C. *A Short Account of Early Muslim Architecture* (Penguin Books, London, 1958)

De Camp, L.S. *Ancient Engineers* (Tandem Books, London, 1977)

Diehl, C. 'Byzantine Art', Ch. 4 of N.H. Baynes and H. St L.B. Moss (eds.),
    Byzantium (Oxford Paperbacks, Oxford, 1961)
Diels, H. *Antike Technik* (Leipzig and Berlin, 1914)
_____ 'Über die von Prokop beschriebene Kunstuhr von Gaza', *Abhandl. d.
    preuss. Akad. Wiss. Berlin* (Teubner, Phil. Hist. Klasse), no. 7 (1917)
Dilke, O.A.W. *The Roman Land Surveyors* (David and Charles, Newton Abbot,
    England, 1971)
al-Dimashqī, *Cosmographie de Chems-ed-Din Abou Abdallah Mohammed
    ed-Dimichqui*, ed. A.F. Mehren (Arabic text, St Petersburg, 1866)
Drachmann, A.G. *The Mechanical Technology of Greek and Roman Antiquity*
    (University of Wisconsin Press, Madison, Wisconsin and Hafner, London,
    1963)
_____ 'Ktesibios, Philon and Heron; a Study in Ancient Pneumatics', *Acta
    Historica Scientiarum Naturalium et Medicinalium*, ed. Biblioteca Universitatis
    Hauiensis, Copenhagen, 4 (1948), pp. 1-197
_____ 'A Note on Ancient Cranes' in *A History of Technology*, ed. Singer *et al.*,
    vol. 2, pp. 658-62
Drover, C.B. 'A Medieval Monastic Water-clock', *Antiquarian Horology*, 1
    (1954), pp. 54-8, 63
Dūrī, A.A. 'Baghdād, EI, vol 1 pp. 894-908
Eliséeff, N. 'Hisn al-Akrād, EI, vol. 3, pp. 503-6
Ellis, M. (ed.) *Water and Wind Mills in Hampshire and the Isle of Wight*
    (Southampton University Industrial Archaeology Group, 1978)
Forbes, R.J. *Notes on the History of Ancient Roads and their Construction* (2nd
    edn, A.M. Hakkert, Amsterdam, 1964)
_____ *Studies in Ancient Technology* (6 vols., Brill, Leiden 1957-8, vol. 1, 2nd
    edn 1964, vol. 2, 2nd edn 1965)
_____ 'Roads and Traffic in the Middle Ages' in *A History of Technology*, ed.
    Singer *et al.*, vol. 2, pp. 524-7
_____ 'Power' in *A History of Technology*, ed. Singer *et al*, vol. 2, pp. 589-622
Freese, S. *Windmills and Millwrighting* (David and Charles, Newton Abbot,
    England, 1957, reprint 1974)
Frontinus, *The Stratagems and the Aqueducts of Rome*, ed. with Latin texts and
    English trans. by Charles E. Bennett and Mary B. McElwain (Loeb Classics,
    Cambridge, Mass. and London, 1925, reprint 1980)
Gille, B. 'Machines' in *A History of Technology*, ed. Singer *et al.*, vol. 2, pp.
    629-57
Gimpel, J. *The Medieval Machine* (Victor Gollancz, London, 1977)
Glick, T.F. *Irrigation and Society in Medieval Valencia* (Harvard University Press,
    Harvard, 1970)
_____ *Islamic and Christian Spain in the Early Middle Ages*, (Princeton
    University Press, Princeton, 1979)
Goblot, H. *Les Qanats; une Technique d'Acquisition de l'Eau* (Mouton Éditeur,
    Paris, 1979)
Hamarneh, S. 'Sugar-cane Plantation and Industry under the Arab Muslims during
    the Middle Ages', *Proceedings of the First International Symposium for the
    History of Arabic Science* (Aleppo University, 1976), p. 221
Hartenberg, R.S. 'Hand Tools', *Encyclopaedia Britannica* (15th edn 1974), pp.
    605-24
Hartner, W. 'Aṣṭurlāb', EI, vol. 1, pp. 722-8
Harvey, J. *English Medieval Architects* (Batsford, London, 1954)
Hassan, A.Y. *Taqī al-Dīn and Arabic Mechanical Engineering*, in Arabic
    (University of Aleppo Press, 1976)
Hellenkemper, H. 'Byzantischer Brückenbau', *Lexicon des Mittelalters* (Artemis

Verlag, Munich and Zurich, 1982), vol. 2, pt. 2, pp. 730-1
Hero of Alexandria, *Pneumatics*, 1st edn B. Woodcroft (Taylor, Walton and
   Maberly, London, 1851), facsimile of Woodcroft, ed. with introduction by
   Marie Boas Hall (Macdonald, London and New York, 1971)
———— *Mechanics*, ed. with French trans. by Carra de Vaux, 'Les Mechaniques ou
   l'Élevateur de Heron d'Alexandrie sur la Version de Qostā ibn Lūqā', *Journal
   Asiatique*, ninth series (1893), vol. 1, pp. 386-472, vol. 2, pp. 152-269,
   450-514
Hill, D.R. *On the Construction of Water-clocks*, Occasional Paper no. 4 (Turner
   and Devereux, London, 1976)
———— 'A Treatise on Machines', *Journal for the History of Arabic Science*, vol. 1,
   no. 1 (Aleppo, 1977) pp. 33-44
———— 'Ḳustā b. Lūḳā', in EI, vol. 5, p. 529
———— *Arabic Water-clocks* (Institute for the History of Arabic Science, Aleppo,
   1981)
Huart, C. and Grohmann, A. 'Kāghad' in EI, vol. 4, pp. 419-20
Hunter, L.C. 'The Living Past in the Appalachians of Europe', *Technology and
   Culture*, vol. 8 (1967) pp. 445-66
Ibn Baṭṭūṭa, Abu 'Abd Allah Muḥammad, *Riḥla*, ed. Karam al-Bustani (Beirut,
   undated)
Ibn Ḥawqal, Abu'l-Qāsim Muḥammad, *Kitāb Ṣūrat al-Arḍ*, ed. J.H. Kramers, 2nd
   edn of vol. 2 of BGA (Brill, Leiden, 1938)
———— *Configuration de la Terre*, French translation by J.H. Kramers and G. Wiet
   (2 vols., Beirut/Paris, vol. 1 1964, vol. 2 1965), page references to Kramer's
   Arabic text are given
Ibn Jubayr, *Riḥla*, 2nd edn M.J. de Goeje (Brill, Leiden, 1907)
Ibn Khurrādadhbih, *Kitāb al-masālik*, extracts in vol. 6 of BGA, ed. M.J. de Goeje
   (Brill, Leiden, 1889)
Ibn al-Nadīm, *Kitāb al-Fihrist*, ed. G. Flugel (2 vols., Leipzeg, 1871-2, reprint 1
   vol. Cairo, 1929-30)
Ibn Sarabiyūn (also called Ibn Serapion) *Kitāb ajā'ib al-aqālīm al-sab'a*, ed. H. von
   Mžik (Vienna, 1929)
Ibn al-Ukhuwwa, Ḍiyā' al-Din Muḥammad, *Ma'alim al-qurba fī aḥkām al-ḥisba*,
   ed. R. Levy with notes and an abridged English trans. (Gibb Memorial Series,
   new series, London, 1938)
al-Idrīsī, Abu 'Abd Allah Muḥammad, *Description de l'Afrique et de l'Espagne*,
   Arabic text with French trans. by R. Dozy and M.J. de Goeje (Brill, Leiden,
   1866)
al-Iṣṭakhrī, Abu Isḥāq Ibrāhīm, *Kitāb al-masālik wa'l-mamālik*, ed. M.G. 'Abd
   al-'Āl al-Hīnī (Cairo, 1961)
al-Jazarī, ibn al-Razzāz, *The Book of Knowledge of Ingenious Mechanical Devices*,
   annotated and trans. by Donald R. Hill (Reidel, Dordecht, 1974)
———— *A Compendium on the Theory and Practice of the Mechanical Arts*, Arabic
   text of the above, ed. A.Y. Hassan (Institute for the History of Arabic Science,
   Aleppo University, 1979)
Kennedy, E.S. 'The Equatorium of Abu al-Salt', *Physics*, Year 12, fasc. 1 (1970),
   p. 80
al-Khuwārizmī, Abu 'Abd Allah Muḥammad, *Liber Mafātīh al-Olūm*, ed. G. van
   Vloten, Arabic text with Latin critical apparatus (Brill, Leiden, 1895)
King, D.A. 'On the Astronomical Tables of the Islamic Middle Ages', *Colloquia
   Copernicana*, 3 (1975), pp. 37-56
———— 'Kibla' in EI, vol. 5, pp. 83-8
Klemm, F. *A History of Western Technology*, trans. Dorothy W. Singer (George
   Allen and Unwin, London, 1959)

van Laere, R. 'Techniques Hydrauliques en Mésopotamie Ancienne', *Orientalia Lovaniensis Periodica*, 11 (University of Leuven Press, 1980) pp. 11-53

Lombton, A. K. S. 'Kanata' in EI, vol. 1, pp. 630 33

Landels, J.G. *Engineering in the Ancient World* (Chatto and Windus, London, 1978)

Le Strange, G. *The Lands of the Eastern Caliphate* (Frank Cass, London, 1st edn 1905, this edn 1966)

Levi-Provençal, E. *Trois Traités Hispaniques de Hisba* (Arabic text — Cairo 1955)

Lewis, B. 'Ḥadjdj' in EI, vol. 3, pp. 37-8

Lloyd, H.A. *Giovanni de' Dondi's horological masterpiece* (Hookwood, Limpsfield, Oxted, Surrey, 1956)

Maddison, F., Scott, B. and Kent, A. 'An Early Medieval Water-clock', *Antiquarian Horology*, 3 (1962), pp. 348-53

Marçais, G. 'Binā'' in EI, vol. 1, pp. 1226-9

Margary, I.D. *Roman Roads in Britain* (John Baker, London, 1967)

al-Mas'ūdī, Abu'l-Ḥasan, *Kitāb al-tanbīh wa'l-ishrāf*, ed. M.J. de Goeje, vol. 8 of BGA (Brill, Leiden, 1894)

Michel, H. *Traité de l'Astrolabe* (Alain Brieux, Paris, 1876)

Mills, A.A. 'Newton's Water Clocks and the Fluid Mechanics of Clepsydrae', *Notes and Records of the Royal Society of London*, vol. 37, no. 1 (1982), pp. 35-61

Molenaar, A. *Water Lifting Devices for Irrigation*, FAO Paper no. 60 (1956)

Moritz, L.A. *Grain Mills and Flour in Classical Antiquity* (Oxford University Press, Oxford, 1958)

Muendel, J. 'Horizontal Water-wheels in Medieval Pistoia', *Technology and Culture*, vol. 15 (1974) pp. 194-225

al-Muqaddasī, Abu 'Abd Allah Shams al-Dīn, *Aḥsān al-taqāsīm fī ma'rifat al-aqālīm*, ed. M.J. de Goeje, vol. 3 of BGA (Brill, Leiden, 1906)

Needham, J. with Wang Ling and Derek de Solla Price, *Heavenly Clockwork* (Cambridge University Press, Cambridge, 1960)

—— *Science and Civilisation in China* (5 vols. to date, continuing, Cambridge University Press, Cambridge, 1954 onwards)

Neugebauer, O. 'The Early History of the Astrolabe' *ISIS*, vol. 40 (1949) pp. 240-56

Pannell, J.P.M. *An Illustrated History of Civil Engineering* (Thomas and Hudson, London, 1964)

Parsons, W.B. *Engineers and Engineering in the Renaissance*, 2nd edn (MIT Press, Cambridge, Mass., 1968)

Pellat, C. 'Baghl' in EI, vol. 1, p. 909

Perkins, J.W. 'Nero's Golden House', *Antiquity*, 30 (1956) pp. 209-19

Philo of Byzantium, 'Le Livre des Appareils Pneumatiques et des Machines Hydrauliques par Philon de Byzance', Arabic text ed. with French translation by Carra de Vaux, *Paris Académie des Inscriptions et Belles Lettres*, 38 (1903), pt. 1

Pingree, D. ''Ilm al-Hay'a' in EI, vol. 3, pp. 1135-8

Poulle, E. 'L'Astrolabe Medieval', *Bibliothèque de l'École de Chartres*, no. 112 (1954), pp. 81-103

—— 'Les Instruments astronomiques de l'Occident latin aux XIᵉ et XIIᵉ siècles', *Cahiers de civilisation medievale*, vol. 15 (1972), pp. 27-40

Pounds, N.J.G. *An Economic History of Medieval Europe* (Longman, London, 1978)

Price, D. de Solla, *Science since Babylon*, enlarged edn (Yale University Press, New Haven and London, 1976)

—— *Gears from the Greeks* (Science History Publications, New York, 1975)

_____ 'Mechanical Water Clocks in the 14th Century ir Fez, Morocco', *Ithaca*, 26, VIII-2 IX (Hermann, Paris, 1962)

al-Qalqashandī, Shihāb al-Dīn, *Ṣubḥ al-a'sha fī sina'āt al-inshā*, ed. M.A.R. Ibrahim (14 vols., Cairo, 1913-20)

al-Qazwīnī, Zakarīya b. Muḥammad, *Athār al-bilād wa akhbār al-'ibād* (Beirut, 1960)

Qudāma, *Kitāb al-Kharāj*, extracts in vol. 6 of BGA, ed. M.J. de Goeje (Brill, Leiden, 1889)

Ragavan, V. 'Yantras or Mechanical Contrivances in Ancient India', *The Indian Institute of Culture*, Basavangudi, Bangalore, Transaction no. 10 (1952)

Richter, G.M.A. 'Ceramics: Medieval' in *A History of Technology*, Singer *et al.*, (eds.), vol. 2, pp. 284-310

Rico y Sinobas, M. (ed.) *Libros del Saber de Astronomia* (5 vols., Aguado, Madrid, 1863)

Robertson, D.S. *Greek and Roman Architecture*, 2nd edn (Cambridge University Press, Cambridge, 1943)

Rohr, R.R.J. *Sundials*, trans. from the French edn of 1965 by G. Godin (University of Toronto Press, Toronto, 1970)

Sabra, A.I. "Ilm al-Ḥisāb' in EI, vol. 3, pp. 1138-41

Sagui, C.L. 'La meunerie de Barbegal (France) et les roues hydrauliques chez les anciens et au moyen age', *ISIS*, vol. 38 (1948), pp. 225-31

Saunders, H.N. *The Astrolabe* (Bude, England, 1971)

Sayili, A. 'Gondēshāpūr', in EI, vol. 2, p. 1120

Schiøler, T. *Roman and Islamic Water-Lifting Wheels* (Odense University Press, Odense, 1973)

Schnitter, N. 'Romische Talsperren', *Antike Welt*, vol. 2 (1978), pp. 25-32

Schreiber, H. *The History of Roads*, trans. S. Thomson (Barrie and Rockcliff, London, 1961)

Shelby, L.R. 'The Role of the Master Mason in Mediaeval English Building', *Speculum*, vol. 39, no. 3 (1964) pp. 387-403

_____ 'The Geometrical Knowledge of Mediaeval Master Masons', *Speculum*, vol. 47 (1972), pp. 395-421

Shirley-Smith, H. *The World's Great Bridges* (Phoenix, London, 1953)

Singer, C., Holmyard, E.J., Hall, A.R. and Williams, T.I. (eds.), *A History of Technology* (7 vols., Oxford University Press, Oxford, 1st edn 1956, this reprint 1979)

Sleeswyk, A.W. 'Vitruvius' Waywiser', *Archives Internationales D'Histoire des Sciences*, vol. 29 (1979)

_____ 'The 13th Century "King Hezekiah" Water-clock', unpublished paper (University of Groningen)

Smith, N.A.F. *A History of Dams* (Peter Davies, London, 1971)

_____ *Man and Water* (Peter Davies, London, 1975)

_____ 'Attitudes to Roman Engineering and the Question of the Inverted Siphon', *History of Technology*, vol. 1 (1976) pp. 45-71

Sourdel, D. 'Barīd' in EI, vol. 1, pp. 1045-6

Spargo, J.W. *Virgil the Necromancer* (Harvard University Press, 1934)

Straub, H. *A History of Civil Engineering*, English translation by E. Rockwell (Leonard Hill, London, 1952)

Symonds, R.W. 'Furniture: Post-Roman' in *A History of Technology*, ed. Singer *et al.*, vol. 2, pp. 221-58

Taccola, Mariano, *Mariano Taccola and his Book De Ingeneis*, ed. F.D. Prager and G. Scaglia (MIT Press, Cambridge, Mass., 1972)

Tait, H. *Clocks in the British Museum* (Trustees of the British Museum, London, 1968)

Terrasse, H. 'Gharnāta' in EI, vol. 2, p. 1019

Theophilus, *On Divers Arts*, trans. from the Latin with introduction and notes by J.G. Hawthorne and C.S. Smith (Dover Publications, New York, 1979)

Thomson, R.H.G. 'The Medieval Artisan' in *A History of Technology*, ed. Singer *et al.*, vol. 2, pp. 383-96

Thorndike, L. 'Invention of the Mechanical Clock about 1271 AD', *Speculum*, 16 (1941), pp. 242-3

Usher, A.P. *A History of Mechanical Inventions*, 2nd edn (Harvard University Press, Cambridge, Mass., 1954)

Vallicrosa, J.M. Millás, *Estudias sobre Azarquiel* (Madrid, 1950)

Verantius, Faustus, *Machinae Novae*, ed. F. Klemm (Heinz Moos, Munich, 1965)

Villard de Honnecourt, *The Album of Villard de Honnecourt*, ed. H.R. Hahnlöser (Schroll, Vienna, 1935)

Vitruvius, *De Architectura*, ed. F. Granger (Loeb Classics, London, 1934, reprint 1970)

Wailes, R. 'A Note on Windmills' in *A History of Technology*, ed. Singer *et al.*, vol. 2, pp. 623-8

Walz, R. 'Beiträge sur altesten Geschichte der altweltlichen Cameliden' in *Actes du 4ᵉ Congres international des Sciences Anthropologiques et Ethnologiques* (Vienna, 1952, publ. 1956)

Watt, W.M. *Companion to the Qur'ān* (George Allen and Unwin, London, 1967)

Webster, R.S., MacAlister, P.R. and Etting, F.M., *The Astrolabe: Some Notes on its History, Construction and Use* (Lake Bluff, Illinois, 1974)

White, L. *Medieval Technology and Social Change* (Oxford University Press Paperback, Oxford, 1964)

––––––– *Medieval Religion and Technology* (University of California Press, 1978)

Wiedemann, E. *Aufsätze zur arabischen Wissenschaftsgeschichte* (2 vols., Olms, Hildesheim, 1970)

––––––– 'Ein Instrument, das die Bewegung von Sonne und Mond dargestellt, nach al-Birūnī', *Der Islam* (1913), vol. 4, pp. 5-13

Wiedemann, E. and Hauser, F. 'Über Vorrichtungen zum Heben von Wasser in der islamischen Welt', *Jahrbuch des Vereins Deitscher Ingenieure*, vol. 8 (1918) pp. 121-54

Wissmann, H. von, and Kussmaul, F. 'Badw' in EI, vol. 1, pp. 872-92

Wulff, H.E. *The Traditional Crafts of Persia* (MIT Press, Cambridge, Mass., 1966, reprinted 1976)

Index

Note: In sorting alphabetically the Arabic article al- is ignored.

Abbasids 8, 21, 22, 232; postal service 86
Abbey of St Sauvere de Vicomte 173
Abbot Hugh of Cluny 99
Abbot Suger of Saint-Denis 99
Abraham ibn Ezra 195
Abu Burda 141
Achaemenids 56, 77
Adana bridge 70
Adda river 72
Aḍud al-Dawla 57, 142, 165
Afghanistan 25; cantilever bridges 61; pontoon bridges 66; suspension bridges 61-2; windmills 173
Agrigento 108
Aḥmad b. al-Ṭayyib 141
Ahwaz: bridge 70; norias 142; waterworks 31, 56-7
Albertus Magnus 206
Alconetar bridge 62
Aleppo 165
Alexander the Great 1, 56, 77
Alexandria 3, 19, 71, 200; colonnades 79; intellectual centre 188, 193
Alfonso X of Castile 194
'Alī b. 'Īsa 193
Alicante 22
Almansa dam 59
Almonacid de Cuba dam 59
Almunecar bridge 30
Amalfi 99
Amid 108
analysis of structures 98-101
Aniene river 53, 104
Antikythera mechanism 186-188, 223, 242
Antioch 3; marble-paved streets 79
Antipater of Thessalonica 158
Anushirwān 164

Apollodorus of Damascus 62
Appius Claudius 81
Aqua Appia 29, 37
Aqua Marcia bridge 69-70
Aqua Trajana 161
aqueduct bridges: Muslim 44; Roman 69-70
aqueducts 17, 28, 29; Constantinople 55; Roman 29-30
Arab conquests 3
Arabia 36, 84
Arabic: diffusion 3; loan words in Spanish 24; translation from Arabic into Latin 194-5; translation into Arabic: from Farsi 3, from Greek 3, 20, from Syriac 202
Arabs 32, 77, 161; see also Islam, Muslims
Aramaic 1
arch: ogival 98-101, origin and diffusion 99; semicircular 99
Archimedes 2, 6, 8, 134, 200, 212, 231; lifting machines 109; screw 132-3
'Archimedes' water-clock 212, 230-2, 235, 237
architects 8; English 109, 112-14
Ardashīr I 56, 77
Arles 161, 163, 164
Armenia 21, 36
artisans 9-10; masons 112-14
Asia Minor 1, 77; camels 84; dams 53; roads 81; water-wheels 162
Assam 61
Assyrians 65
astrolabe: construction, history and use 190-5; in surveying 120-1
astronomy: Ptolemaic 185-6; spherical 183-4
Athens 78, 79: overshot wheel 159; public buildings 104;
Atrush river 28

Augsburg 167
Augustus 103, 104-5
automata 199-221; antiquity 200;
    components and techniques
    207-13; examples 213-21; in
    water-clocks 230-2, 235, 236,
    237; India 224; legends 199;
    musical 204-5, 211, 237
Avignon bridge 69, 71

Baalbek 109
Babylon: bricks 103; clepsydra 224
Baghdad 4, 8, 22-3, 25; automata
    205-6; paper-mills 169; pontoon
    bridges 66; population 4, 166
al-Balādhurī 24, 141
Band-i-Amir dam 57-8, 142
Band-i-Qassār dam 170
Banū Mūsà 8, 10, 204, 205, 206,
    208, 220; *Book of Ingenious
    Devices* 202; control systems 212;
    horizontal water-wheel 159-60,
    162; lamp 215-16; *Measurement
    of Plane and Spherical Figures*
    202; trick vessel 216-19; wind-
    wheel 172
Barada river 21
Barbegal 161, 163-4
Barcelona 241
basin irrigation 18
Basra 22, 24; foundation 25; norias
    141; tidal mills 166
baths: Baghdad 31; Damascus 31;
    Tiberias 44; 'Turkish' 44
Bavian dam 28
Bayonne 170
Bazacle dam 60, 168
Belisarius 161
Bilāl b. Abī Burda 141
Bin canal 24
Al-Bīrūnī 8, 120, 124; geared
    calendar 188-90, 242;
    water-powered trip-hammers 169
Bolsena 142
bricks 102-4
bridges 61-74; arch 67-9; beam 63-4;
    cantilever 61; drawbridge 63;
    foundations 72-4; pontoon 65-7;
    suspension 61-2; truss 62
Brindisi 81
Britain 32; bricks 103; Cistercians
    171; grist milling 166-7; windmills
    174-5
Bukhara 26

Burma 61
Byzantine Empire 2, 32; architecture
    98, 101; bricks 103; grist milling
    164; trade with Islam 86;
    water-clocks 230, 232; water
    supply 32
Byzantine siphon 7, 150
Byzantium 3, 27, 163

Caesar 64, 73
Cairo 87
caisson 73-4
calibration of orifices 213, 221
Caliph Hishām b. 'Abd al-Malik 3
Caliph al-Ma'mūn 202
Caliph al-Muqtadir 205
Caliph 'Umar I 25, 173
camels: Bactrian 84; dromedary 84,
    85
Campanus 197
cams 169, 170
canals 17, 22; Basra 25; Nineveh 28;
    Sughd (Transoxiana) 26; *see also*
    canals by name
Canatello 80
candle clocks: al-Jazari 238-9; *Libros
    del Saber* 240
Cantabrian mountains 167
Canterbury 82, 173; cathedral 108,
    waterworks 206
Capua 81
casting metals 108
Catalonia 171-2
Cesena dam 59
Champagne 171
channels 209; terracotta 27
Charlemagne 232
Chartres cathedral 108
Château-Narbonnais dam 168
Cherchel bridge 70
Chesterfield 108
China: camels 84; cantilever bridges
    61; noria 140; segmental arch
    bridge 68-9; suspension bridges
    61-2; water-clocks 226-7;
    water-powered trip-hammers 169;
    water-wheels 160-2; windmills
    173
Chollerford Bridge 163
chorobates 117
Cicilian gates 77
Cismone river 62
Cistercians: timekeeping 241; use of
    water power 171

cisterns 27; Constantinople 32; Qayrawan 144-5; Rome 40
Claudius 70
Cleopatra 1
clepsydra 224-5; steelyard 226-7, 233-4; *see also* water-clocks
clocks: *see* candle clocks, compartmented cylinder clock, mechanical clocks, water-clocks
coffer-dam 73-4
Cologne 242
colonnades 79
compartmented cylinder clock 240-1
concrete 53, 70, 83, 105; water-resistant 37, 73
conscription of artisians 9
Constantinople 32; dams 55; Haghia Sophia 101; 'Throne of Solomon' 205; water supply 32
Coptic 1
Cordoba 233; arch bridge 69; dam 165
corn milling: *see* grist milling
Cornalvo dam 54
Corneto 80
Cotswold region 105
cranes 110-12
crank 132, 143
Crete 157, 162
Ctesibius 7, 213; organ 200-1; pump 142-3; water-clocks 227-8
Ctesiphon 20
cultural divisions 1-4
Cyrene 79
Cyrus 77

Damascus: fountains 44, 206; 'Gate of the Hours' 232; Ghuta (oasis) 21, 42; Ridwan's water-clock 234
dams 47-60; antiquity 50-2; Byzantium 55; design and construction 47-9; diversion 29, 57; Europe 59-60; flood protection 54; Iran 56-7; irrigation 19, 24; Muslim 57-8; Roman 52-5; water power 165
Dante 174, 206, 241
Danube river 62, 81
Dara dam 55
Daurade dam 168
Delphi 79
al-Dimashqi 173
Diocletian 54, 82
Diyala river 24

Diyar Bakr 146, 203
Dizful bridge 70
Dome of the Rock 101
domes 101
Domesday Book 166-7
donkey 84
Dover clock 245
Duc Phillipe, Count of Artois 207
Dujayl river 70
Durance river 163

Ecbatana 77
Egypt 25, 27, 32, 77, 117; arches 99; automata 200; basin irrigation 18; camels 84; clepsydra 224; Great Pyramid 116; irrigation 17, 19-20; mills at Bilbays 164; paper-mills 169; porphry 105; postal service 87; qanats 36; saqiya 138-9; screw 133; windmills 173
Elba 105
Elche 22
Elis 79
engineers 5-9; Abbasid 24; Assyrian 28; Greek 29; Islamic 202; knowledge and training of 7-8; status 8-9
England: grist milling 167; mechanical clock 245; windmills 176
Ephesus 77, 79; temple of Diana 110
equatoria 195-8; construction 195-7; early treatises 197
Erechtheum 104
Etruscans 52, 117
Euphrates 25, 27, 77, 81; pontoon bridges 66; river traffic 85; shifts in course 22-4; ship-mills 165
Europe 129, 174; communications 87-9; milling 161; prehistoric 27; technical achievements 5; water-clocks 241-2; water-wheels 162

Farsi 3
feed-back control 230
Fez: water-clocks 240; water mills 165
floats 213; *see also tarjahār*
Florence 72
forges 170-1
fountains 44; Damascus 206; al-Jazari 220-1;Rome 40
Fourth Crusade 32

France 32, 87, 133; aqueduct bridges 70 ; bricks 103; dams 53, 60; grist milling 167; mechanical clock 245; water-clocks 241-2
Francesco di Giorgio Martini 143
fulling mills 170-1
Fustat 66

Garonne river 60; dams 168
gears 208; epicyclic 203, 208; segmental 203, 208
Generalife water gardens 31-2
Geometria incerti auctoris 122
geometry: Archimedes 6; medieval masons 6-7
Gerard of Cremona 202
Gerasia 79
Gerbert d'Aurillac (later Pope Silvester II) 194
Germain-en-Laye 142
Germany 32, 87; Cistercians 171; grist milling 167; water-driven trip-hammers 170
Gezer 28
Giovanni de' Dondi 148; astronomical clock 245
Glanum 53
Godalming 108
Gondeshapur 3
Gothic architecture 98-101
granite 105
Great Swamp of Iraq 23-4
Greece 27; 'sacred roads' 78-9
Greek 1, 2; translations into Arabic 3, 203
Greeks: architecture 98, 101, 108; clepsydra 224-5; masonry buildings 104-5; surveying 117; use of iron in buildings 107-8
grist milling 4, 159, 162-9; Byzantine Empire 163-4; Europe 166-9; Islam 164-6; Roman Empire 163-4; Sasanid Empire 164; *see also* millstones, water power, water-wheels
groma 119-20
Guadalquivir river 69, river traffic 85

Habash al-Hāsib 124
Haditha 165
Hadrian 55, 70
Hagia Sophia 101
Hama 139, 140, 141
Hamdanid dynasty 165

Hampshire 176
Harbaqa dam 54
Harran ?
Harūn al-Rashīd 232
Hellenistic Age 1-2; architecture 98; automata 200, 214-15; fine technology 188; irrigation 17, 19-20; norias 140; technical achievements 5; water-clocks 227-32; water-wheels 160-1
Hellespont 65
Helmand river 66
Herat 166
Hermann le Boiteux 194
Hermann of Dalmatia 195
Hero of Alexandria 2, 6, 8, 202, 206, 208, 212; aeolipile 215; lifting machines 109; pump 142; surveying 120, 122; translation into Arabic 188; wind-wheel 172
Herod 79
Herodotus 28, 65, 77
Hilla 23, 66
Himalayan region 61
Himyarites 51
Hippodamus of Miletus 117
hodometer *see* waywiser
Homs dam 54
horse 85
Hull 103
hydraulic cement 69, 72, 83, 106

Iberian peninsula 1, 19, 20, 22; dams 53; irrigation 19, 21; saqiya 139; water mills 167; *see also* Muslim irrigation
Ibiza 135
Ibn al-'Āsakir 170
Ibn al-'Awwām 139
Ibn al-Balkhī 170
Ibn Hawqal 61, 165-6
Ibn al-Haytham 232
Ibn Jubayr 43, 44, 66, 71, 142, 166, 206
Ibn Khaldūn 202-3
Ibn Khurradadhbih 86
Ibn al-Nadīm 231
Ibn al-Saffār 120
Ibn al-Samh 197
Ibn Sīda 139
Ibn Tūlūn 71
Ibn al-Ukhuwwa 44
iced water dispensers 44
Idhaj bridge 71

al-Idrīsī 30, 44, 66, 69
India 18, 77; alignment of mosques
123; automata 225; noria 140;
ogival arch 99; suspension bridges
61-2; terrace irrigation 18;
water-raising machines 131;
windmills 173
instruments 183-98; astrolabes 190-5;
calendars 186-90; equatoria
195-7; surveying 117-20
inverted siphon 27, 28; Lyon 30, 38
Iran: arches 99; brick making 102-3;
camels 84; dams 57-8; grist
milling 163; horizontal
water-wheel 160-2; paper-mills
169; postal service 86; qanats 31;
water-clocks 230
Iraq 25, 77; bricks 104; camels 84;
canals 10; dams 57; flour supply
165-6; irrigation 19-20; norias
141; river traffic 85
Ireland 159
irrigation: diffusion 20-1; Egypt
17-19, 20; Hellenistic 20;
Mesopotamia 17-19; Nabateans
19, 20; Romans 21; Sasānid 20,
22-4; Spain 21-2; types 18-19;
Visigoths 21; Yemen 19; *see also*
Muslim irrigation
Irthing river 163
Irtysh river 85
Isfahan 71, 135
Islam 1, 2, 32; commerce 86;
connotation in this book 3-4; grist
milling 163; mobility 85; technical
achievements 5; *see also* Muslims
Islamic land communications 84-7;
animals 84-5; commerce 86;
pilgrimage to Mecca 85-6; postal
service 86-7
Issus 77
Istakhr *see* Persepolis
al-Iṣṭakhrī 25, 66, 71, 86, 142;
attitude to water power 164;
windmills 172-3
Italy 18, 32, 87; aqueduct bridges 70;
dams 59-60; mechanical clocks
245; terrace irrigation 18; roads
81

Jaen 165
Jativa 169
al-Jazarī 7, 8, 10, 170, 207, 208, 210,
211-12, 213, 230; alternating

fountain 220-1; *Book of
Knowledge of Ingenious
Mechanical Devices* 203-5; palace
doors 108; phlebotomy devices
219-20; water-clocks 236-8;
water-raising machines 146-52
al-Jazira *see* Upper Mesopotamia
Jerusalem: Dome of the Rock 101;
water supply 27-8
Jerwan bridge 27-8
Jocelin of Brakelond 173
John of Salisbury 206
Jordan 19
Justinian 32, 70

Karun river 31, 56
Kashmir 61
Kasserine dam 54
Khabur river 142
al-Khalīlī 124
Khalis canal 24
al-Khāzinī 232; steelyard clepsydras
233-4
Khosr river 28, 50
Khurasan 25
al-Khuwārizmī, Abu 'Abd Allah:
Keys of the Sciences 109, 203
al-Khuwārizmī, Muhammad b. Mūsà
193
al-Kindī 201
Kish 26
Kitāb al-Hāwī 42, 117, 131, 139,
144, 146
Knossos 27
Kufa 23
Kur river 56, 57, 142, 165, 170

Laodicea 77
lathe 12
Lebanon 18, 109
Leck river 167
Leonardo da Vinci 122; conical
valves 212; pump 143
Leptis Magna 20; dams 54
Li Chhun 68
Libros del Saber 194, 197; clocks 240
Liutprand of Cremona 205
Llobet or Lupitus 194

machines 6
Magyars 87
Majjana 166
Mamlūks 87
marble 104-5

Marib dam 19; hydraulic works 51
Marv irrigation works 25, 42
Marzabotto 80
master masons: function and training
  112-14; use of geometry 6-7
al-Mas'ūdī 2; windmills 173
materials of construction 102-9;
  bricks 102-4; cobwork 107;
  metals 107-9; mortar 105-6; stone
  104-5; timber 106-7
Mecca: pilgrimage 85-6; qibla 123-4
mechanical clock 223, 242-5;
  ancestry 242; escapement 243-4;
  'striking train' 244
mechanistic philosophy 199
Megiddo 28
mercury 203, 205, 240
Merida 165
Merida bridge 70
Mesopotamia (ancient) 27, 69; brick
  making 102; irrigation 17; *see also*
  Upper Mesopotamia
Messahalla 193
Metz 142
Midianites 84
Milan: dam 59; fulling mills 170
Miletus 117
millstones 163, 164, 166
Mithridates 158
Monastery of Ripoll 122, 194;
  water-clock treatise 241
Mongols 26, 87
Monte Cassino 99
Moselle river 161
mosques 101, 108
Mosul 28; ship-mills 165; trade 86
moving heavy loads 109
Muḥammad al-Sa'ātī 234, 236, 237,
  240
muḥtasib (Islamic official) 9
mules 85
al-Muqaddasī 31, 56, 57, 141, 206,
  232; attitude to water power 164
al-Murādī 148, 203, 208, 233, 242;
  water-clocks 233
Murcia 22; dams 58; ship-mills 165
Murghab river 25; water mill 161
Muslim hydraulics 42-5;
  administration 42
Muslim irrigation 17; allocation of
  water 42-3; Central Asia 25;
  Egypt 25; Iran 25; Iraq 20, 22,
  24, 25, Basra 24-5; Spain 22, 25;
  Upper Mesopotamia 25

Muslims 62, 87; architecture 98, 108;
  bridges 71; exploitation of water
  power 133, 160, trigonometry and
  astronomy 186; *see also* Islam
Mu'tamid b. 'Abbād 193

Nabateans 19, 52, 77; wadi irrigation
  19, 20
Nahr Ma'qil canal 25
Nahr Sura canal 23
Nahr al-Ubulla canal 25
Nahrawan canal 20, 24
Nasaf 26
Naṣīr al-Dīn 204
Naviglio Grande canal 59
Naxos 105
al-Nayrīzī 124
Negev 52, 54; wadi irrigation 19
Nepal 61
Nero 53; 'Golden House' 201-2, 205
Nestorian Christians 3
Nihawand 86
Nile 27, 43, 63, 164; river traffic 85
Nilometer 43-4
Nîmes 70
Nineveh 28, 77; hydraulic works 50
Nishapur 164
Nisibis 77
noria *see* water-raising machines
Norman architecture 98
Normandy 173
North Africa 21, 25; alignment of
  mosques 123; aqueduct bridges
  70; camels 84, 85; cobwork 107;
  dams 53, 54; irrigation 20;
  paper-mills 169; qanats 35, 36
Norwich cathedral clock 245
Novelda 22
Nūr al-Dīn Maḥmud b. Zenki 234

Oleggio 59
Orontes river 54, 139
Örükaya dam 55
Oseney Abbey 173
Ovdat 52

Palermo 165
Palestine 18, 20, 25, 84; Nabatean
  civilisation 19; terrace irrigation
  18; water supply 27-8
Palladio 62
Palmyra 20, 77; colonnades 79; qanat
  36
Pantheon 101; bronze doors 108

paper-mills 169
Pappos 2
Parnassus 78
Partheon 104, 107, 109
Parthians 77
pawl 135
perennial irrigation 18
Pergamon: paved streets 79; water
  supply 28
perpetual motion 205
Persepolis 3, 56, 57, 77, 142
Peterborough cathedral 111-12
Petra 77
Philo 134, 148, 202, 206, 208, 212,
  231; automaton 214-15;
  Pneumatics 201; translation into
  Arabic 188
pilgrimages 85-6, 87-8
pipes: manufacture 209; Knossos 27;
  Rome 40
Piraeus 79, 117
Pistoia 162
Pliny the Elder 11
Po river 81; river traffic 88
Pompeii 11
Pons Aemilius bridge 69
Pons Fabricius bridge 69
Pons Sublicus bridge 64
Pont du Gard bridge 70
Pont Saint-Martin bridge 69
Ponte del Castelvecchio bridge 72
Ponte Vecchio bridge 72
Pope Celestine III 173
postal services: Iran 77, 86; Islam
  86-7
pre-Columbian America 18
Prince Bhoja 225
Procopius 55, 232
Propylaea 104, 108
Proserpina dam 53-4
Ptolemy (astronomer) 2, 185, 186,
  193; translation into Arabic 188;
  translation into Latin 195
Pul-i-Bulaiti dam 57
Pul-i-Kaisar bridge 56
pumps *see* water-raising machines
Pyrenees 167

Qal'at Ja'bar 165
al-Qalqashandī 204
qanats 17, 20, 29, 33-6, 44;
  construction 33-5; productivity 35
Qatr 23
al-Qazwīnī 71

qibla 123-4, 186
quantity surveying 42
Qudāma 86
Qurna 23
Qustā b. Lūqā 201

Raqqa 165
Ra's al-'Ayn 142
Ravenna 101; piled foundations
  106-7
Raymond of Marseilles 195
Rayy 36
reservoirs 27, 29, 211; behind dams
  59; Constantinople 32
Rhine: Caesar's bridge 64, 73; river
  traffic 88
Rhodes 186
Rhône 164
Rialto 66
rice-husking mills 171
Richard of Wallingford 245
Ridwān 212, 230, 231, 232, 233;
  treatise on water-clock 234-6
ritual ablution houses 44
river traffic: Europe 88; Islam 85
roads: Carthage 80; Etruscan 80;
  Greece 78-9; Iran 77-8, Royal
  Road 177; medieval Europe 88-9;
  paving 79, 80, 82-3, 87, 88-9; *see
  also* Roman roads
Robert of Chester 195
Robertus Anglicus 243, 245
Rochester 82
Roman Empire 1-2, 32, 87;
  anaphoric water-clock 229; bricks
  103; dams 52-5; grist milling 163;
  standard weights and measures
  122; standardisation of
  components 40; technical
  achievements 5; tools 11-12
Roman roads 80-4; construction
  81-4; planning 80-1
Roman water supply 29, 36-41;
  aqueducts 37-8; distribution
  system 39-40
Romans 1, 29, 44, 52; arch bridges
  69-70; architecture 98, 101;
  bricks 103; clepsydra 224-5; dams
  52-5; irrigation 20; public works
  84; pumps 142; roads 80-4;
  surveying 117, 119-20; timber
  bridges 64
Romanesque architecture 98, 101,
  108

Rome 20, 29, 53, 81, 201, 225;
armament workers 9; bridges 64;
masonry buildings 104-5, Nero's
'Golden House' 201-2; overshot
wheel 159; pilgrimages to 87;
ship-mills 161; Trajan's column
62; water-wheels 161
Rouen 170

Saalburg 163
Sabeans 51
Sabratha 20
Sadd al-Kafra dam 50
St Albans Abbey 245
Sakians 84
Saladin 204
Salisbury cathedral 111-12
Salzburg 229
Samarkand 26, 44; paper-mills 169;
paved streets 87
Samos 28-9
San Vitale church 101
Sanja bridge 69
Saône river 164
saqiya *see* water-raising machines
Saragossa 59; ship-mills 165
Sardis 77
Sargon II 35
Sasānid Iran 2, 3; bridges 70; grist
milling 164; intellectual activity
2-3; irrigation 17, 19, 22-3; roads
77-8; technical achievements 5;
water mill 161
Savio river 59
Savoy 61
Sayhan river 70
Schmidmulen 170
screw *see* water-raising machines
Scythians 84
Sebokht, Severus 193
segmental gears 147
Segovia bridge 30, 70
Seine 88
Seleucia 79
Seleucids 2
Seleucus 77
Selinus 109
Senchus Mor 159
Sennacherib 28; hydraulic works 50-1
Seville: bricks 104; standard measures
122
shādūf *see* water-raising machines
Shatt a-'Arab river 24-5
sheet lead for roofs 108

Shiraz 57, 142
Shustar: bridges 70; dams 56-7
Silchester 142
siphon: bent-tube 209; concentric
209; double concentric 209-10;
see also inverted siphon
Solomon 27
Spain 21-2, 25, 44, 171-2; allocation
of irrigation water 43; bricks 103;
dams 53-4, 58-9; grist milling
167; irrigation 22; mines 132-3;
paper-mills 169
Sparta 79
stone 104-5
Strabo 159
Su Sung's clock 203, 226, 242
Suetonius 201
sugar-mills 170
Sughd province 26; river traffic 85
Sultan Baybars I 87
sundial 224
surveying 118-24; centuriation 119;
triangulation 122; *see also*
instruments
surveyors 10; *agrimensores* 119, 122
Susa 77
Syria 2, 18, 25, 32, 36, 77; bricks
102; camels 84; dams 54-5;
masonry 105; ogival arches 99;
paper-mills 169; postal service 87;
terrace irrigation 18
Syriac 1; translations into Arabic
from 3, 202

Tab river 61; bridge 70-1
al-Ṭabarī 202
Tabaristan 164
Taccola, Mariano 7-8, 73; pump 143
Tacitus 201
Tagus river 69, 232
Takrit 24
Tangier 36
tanks 27, 211
Taqī al-Dīn 152
tarjahār 43, 213; in water-clocks
237-8
Tarragona bridge 70
Tarsus 77
Taunus range 163
Taurus mountains 61
Teheran 35, 36
temple of Zeus Olympios 104
terrace irrigation 18
Tewkesbury Abbey 111-12

Thailand 61
Theocritus 2
Theon of Alexandria 193
Theophilus 10
Theorica planetarum 197
Tiber 64, 69, 104; ship-mills 164
Tibet 61
Tiflis 165
Tigris 20, 25, 27, 28, 50, 57; 'mill power' 164; pontoon bridges 66; river traffic 85; shifts in course 22-4; ship-mills 165-6
Timur 26, 87
tipping-buckets 212-13, 221
Tivoli 104
Tlemcen 165
Toledo: bridge 30; noria 140; translations from Arabic in 194-5; water-clocks 232-3
tools 11-13; *see also* lathe
Toulouse 168
Tournus 164
Trajan 20, 52
Transoxiana *see* Sughd
tread-wheels: operating chain-of-pots 135; operating screw 132; operating tympanum 134-5; operating winches 111-12
Trent river 64
Trezzo bridge 72
trip-hammers 169, 170
Tuna al-Gabal 138
tunnels: Samos 29; *sinnor* 28
Turkey 25; dams 55
tympanum *see* water-raising machines

Upper Mesopotamia: camels 84; dams 55; granary for Iraq; millstones 166; palace of Amid 110; water-raising 151
Urartu 35
Utica 102
Uzaym dam 24, 57

Valencia 22; allocation of irrigation water 43; dams 58
Valerian 56
valves: clack- 150; conical 212, 217-18, 230; plate- 212
Verantius, Faustus 62
Verona 72
Vespasian 69
Via Appia 81
Via Flaminia 81

de Vigevano, Guido 7, 8
Vikings 87
Villard de Honnecourt 61, 206
Virgil 206
Visconti, Bernabò 72
Visigothic Spain 20-1
Vitruvius, 7, 8, 37-9, 73, 200-1; lifting machines 109-11; list of quarries 104; specifications: bricks 102-3, sand 105, timber 106-7; surveying instruments 117; technical knowledge of 41; water-clocks 227-30; water-raising machines 128, 131-4, 140, 142; water-wheels 159, 160, 161; waywiser 122-3
Volga 85

wadi irrigation 19
Wasit 23
water-clocks 232-42; anaphoric 229; diffusion 231-2, 242; European 241-2; Gaza 232; Hellenistic 227-32; India 225; Islamic 232-41; *Libros del Saber* 240; origins 224-5
water mills: by bridge piers 165, 168-9; gears 157; ownership 164; ship-mills 161, 164, 165, 166, 167, 168; tidal 166, 167; under dams 165, 167, 168; *see also* grist milling, millstones, water power, water-wheels
water power for industrial uses 169-72; diffusion 171-2; *see also* grist milling, millstones, water mills, water-wheels
water-raising machines 127-52; al-Jazari 146-52, twin-cylinder pump 149-52; noria 21, 139-42, 160, construction 139, diffusion 140-1; output 144-6; pumps 7, 142-3, weight-driven 205; sāqiya 135-9, construction 135-8, diffusion 138-9, water-driven 148; screw 131-3; shādūf 130-1; spiral scoop-wheel 138; tympanum 133-4; windlass 128
water supply 26-33; Assyria 28; Byzantine 32; Crete 27; Europe 32-3; Greece 27, 28; Muslim 30-1; Palestine 27-8; *see also* Muslim hydraulics, Roman water supply

water-wheels 31, 155-62; horizontal
157-8, 167; overshot 156-7;
Pelton wheel 211, scoop wheel
210-11; undershot 155-6
Watling Street 82
waywiser 122-3
Weedley 173
Willowford Bridge 163
windlass *see* water-raising machines
windmill 172-7; horizontal axle
173-7; vertical axle 172-3, 176
'Windmill Psalter' 173

Xerxes 65

Yazid river 148
Yemen 19; allocation of irrigation
water 19; paper mills 169; wadi
irrigation 19; *see also* Marib dam
York minster 108
Ypres 173

Zagros mountains 61
Zarafshan river 26
al-Zarqālī (Azarquiel, Arzarchel)
193; treatise on equatoria 197
Zenobia 77
Zeugma 77